The ABCs
of
IP Addressing

OTHER AUERBACH PUBLICATIONS

ABCs of IP Addressing
Gilbert Held
ISBN: 0-8493-1144-6

Application Servers for E-Business
Lisa M. Lindgren
ISBN: 0-8493-0827-5

Architectures for e-Business
Sanjiv Purba, Editor
ISBN: 0-8493-1161-6

A Technical Guide to IPSec Virtual Private Networks
James S. Tiller
ISBN: 0-8493-0876-3

Building an Information Security Awareness Program
Mark B. Desman
ISBN: 0-8493-0116-5

Computer Telephony Integration
William Yarberry, Jr.
ISBN: 0-8493-9995-5

Cyber Crime Field Handbook
Bruce Middleton
ISBN: 0-8493-1192-6

Enterprise Systems Architectures
Mark Goodyear, Editor
ISBN: 0-8493-9836-3

Enterprise Systems Integration, 2nd Edition
Judith Myerson
ISBN: 0-8493-1149-7

Information Security Architecture
Jan Killmeyer Tudor
ISBN: 0-8493-9988-2

Information Security Management Handbook, 4th Edition, Volume 2
Harold F. Tipton and Micki Krause, Editors
ISBN: 0-8493-0800-3

Information Security Management Handbook, 4th Edition, Volume 3
Harold F. Tipton and Micki Krause, Editors
ISBN: 0-8493-1127-6

Information Security Policies, Procedures, and Standards: Guidelines for Effective Information Security
Thomas Peltier
ISBN: 0-8493-1137-3

Information Security Risk Analysis
Thomas Peltier
ISBN: 0-8493-0880-1

Information Technology Control and Audit
Frederick Gallegos, Sandra Allen-Senft, and Daniel P. Manson
ISBN: 0-8493-9994-7

Integrating ERP, CRM, Supply Chain Management, and Smart Materials
Dimitris N. Chorafas
ISBN: 0-8493-1076-8

New Directions in Internet Management
Sanjiv Purba, Editor
ISBN: 0-8493-1160-8

New Directions in Project Management
Paul C. Tinnirello, Editor
ISBN: 0-8493-1190-X

Oracle Internals: Tips, Tricks, and Techniques for DBAs
Donald K. Burleson, Editor
ISBN: 0-8493-1139-X

Practical Guide to Security Engineering and Information Assurance
Debra Herrmann
ISBN: 0-8493-1163-2

TCP/IP Professional Reference Guide
Gilbert Held
ISBN: 0-8493-0824-0

Roadmap to the e-Factory
Alex N. Beavers, Jr.
ISBN: 0-8493-0099-1

Securing E-Business Applications and Communications
Jonathan S. Held
John R. Bowers
ISBN: 0-8493-0963-8

The ABCs
of
IP Addressing

GILBERT HELD

CRC Press
Taylor & Francis Group
Boca Raton London New York

CRC Press is an imprint of the
Taylor & Francis Group, an **informa** business

AN AUERBACH BOOK

Library of Congress Cataloging-in-Publication Data

Held, Gilbert, 1943-
    The ABCs of IP addressing / Gilbert Held.
        p. cm.
    Includes index.
    ISBN 0-8493-1144-6 (alk. paper)
        1. TCP/IP (Computer network protocol) 2. Internet addresses. 3. Directory services
(Computer network technology) I. Title.

TK105.585 H44694 2001
004.6'2—dc21   2001046273

Visit the Auerbach Web site at www.auerbach-publications.com

CRC Press
Taylor & Francis Group
6000 Broken Sound Parkway NW, Suite 300
Boca Raton, FL 33487-2742

First issued in hardback 2017

© 2002 by CRC Press LLC
CRC Press is an imprint of Taylor & Francis Group, an Informa business

No claim to original U.S. Government works

ISBN 13: 978-1-138-47242-6 (hbk)
ISBN 13: 978-0-8493-1144-4 (pbk)

Library of Congress Card Number 2001046273

Dedication

Teaching represents a learning process for the instructor. For this opportunity, which has been provided to me over the past decade, I would be remiss if I did not acknowledge the students of Georgia College and State University who provide a learning laboratory for the presentation of technical information. I also thank Dr. Harry Glover for providing me with the opportunity to teach graduate school at Georgia College and State University.

Contents

Preface

Today we work and play in an electronic-based world. This world is rapidly becoming Internet-based, with tens of millions of homes, millions of businesses, and, within a short period of time, possibly hundreds of millions of mobile professions accessing the literal mother of all networks. Adding fuel to the explosion in the use of the Internet is the development of smart appliances. Perhaps by the time you read this book you might be able to power up your computer at work and initiate a connection to the refrigerator in your home. Using an appliance application program, you might click on an icon on your screen to turn on the light in your refrigerator and, through the use of another icon, control the use of a miniature video camera that allows you to scan the contents of the shelves. Quickly clicking on the entries on a food list displayed as a pop-up menu, you can choose items to pick up on the way home and then print the list or download it into your personal digital assistant (PDA).

If the previous description of the use of a smart appliance appears to be futuristic, interestingly several appliance manufacturers were working on intelligent refrigerators when this book was prepared. However, what might not be readily apparent is the fact that one of the constraints affecting the ability to add hundreds of millions of smart appliances to the Internet as well as mobile phones, PDAs, and other devices is the ability to obtain and use addresses required to access such devices.

The ability to route data across the Internet is based on the use of what is referred to as an Internet Protocol (IP) address, which is the focus of this book. Every device connected to the Internet requires a unique IP address to ensure that data can reach each device. For many Internet-related operations, such as the use of a browser, we usually do not need to be concerned with IP addressing. However, if we need to configure a workstation, server, or even a smart appliance to operate on a network, we must either configure several IP addresses for each device or depend on a network manager or LAN administrator to perform the required configuration. Similarly, if we are installing, modifying, or expanding an IP network, we must carefully consider many aspects associated with IP addresses to include the role of special IP addresses, subnetting, classless inter-domain routing, network address translation, the use of IPv4 versus IPv6, and numerous other factors associated with designing a network architecture, all topics covered in this book.

Most readers should be familiar with the entry of near-English identifiers, more formally referred to as Uniform Resource Locators (URLs), into a browser to access a particular Web page. However, all routing on the Internet occurs via the use of IP addresses. Thus, transparent to most Internet users is the Domain Name Service that automatically translates URLs into IP addresses which enable our Web page request to be routed to an appropriate location on the Internet. While the Domain Name System (DNS) operates transparently to most Internet users, they must configure their devices to locate a DNS server. Thus, the effective use of the Internet requires users to understand a variety of IP addressing information as well as the entry of IP addresses into many device configuration screens.

The importance of IP addressing was the rationale for writing this book. Over a period of approximately ten years, I noted that one of the key problems affecting many Internet users, ranging from individual professionals to network managers and administrators, was a lack of knowledge concerning IP addressing. While many times it is possible to design a network or configure a PC to operate on a network without a detailed knowledge of IP addressing, the end results might not be very efficient. Thus, the purpose of this book is to provide readers with both detailed information concerning IP addressing as well as a comprehensive reference to this topic, which should facilitate the design of networks as well as the configuration and operation of equipment. By including a separate chapter covering network tools, you will also be prepared for testing and troubleshooting operations that can facilitate the isolation and correction of network-related problems.

As an author, I value reader comments. As you read each chapter, I encourage you to create a list of comments and questions concerning this book. Tell me if there are topics that you feel require additional elaboration, topics that should be reduced in coverage, or if you have other comments. Through reader feedback it becomes possible to tailor revisions for a new edition to better meet the expectations of readers. You may contact me either through my publisher or directly via email at gil_held@yahoo.com.

Gilbert Held
Macon, Georgia

Acknowledgments

One of the unfair facts of life is that the name of an author is prominently displayed on the cover of a book while the names of the team that made the book possible are commonly hidden from view. As an author, long ago I recognized the important role of the team of professionals who are required to convert a book proposal into a published book. Thus, I would be remiss if I did not acknowledge the team whose efforts resulted in the publication of this book.

The decision to publish a book is obviously an important role in the eventual publication of the book. While it may appear to represent a simple binary decision, in actuality the process is quite involved and requires careful consideration of the proposed topic and an examination of an economic model because publishers are in business to make a return on their investment. I would like to once again thank Richard O'Hanley at Auerbach Publications and CRC Press LLC for supporting another one of my book ideas.

As an old-fashioned author who travels the globe lecturing and touring interesting sites, I long ago came to the conclusion that it was easier to use pen and paper than to depend on the use of electrical outlet adapters that often would not quite allow my laptop or notebook to be recharged from very interesting wall outlets. Thus, I am highly dependent on exceptional typists who not only understand my longhand writing, but also have the ability to convert my drawings into professional illustrations. Thus, once again I am indebted to Mrs. Linda Hayes and Mrs. Susan Corbitt for their fine efforts in preparing a professional manuscript.

Once a manuscript arrives at a publisher it is checked, rechecked, edited, and converted into page proofs. A cover is designed and a description of the highlights of the book is created for the back cover. The book must be registered with the Library of Congress, sales personnel must market the book, and numerous other details must be addressed. Thus, a behind-the-scenes level of activity is required to create and market a book. Recognizing this effort, I would like to thank those persons at Auerbach and CRC Press whose efforts contributed to this book.

   Last but not least, the development of any book represents a time-consuming effort that requires working during numerous weekends and evenings. Once again I appreciate the understanding of my wife, Beverly, for her willingness to recognize my need to work beyond the normal work week during the preparation of this book.

Chapter 1

Introduction

In the preface to this book, I briefly mentioned a few of the reasons behind the need to become knowledgeable concerning IP addressing. As an introduction, in this chapter we will expand our knowledge of the need for understanding IP addressing as well as focus attention on reviewing the manner by which the Internet operates. In doing so we will note the use of Uniform Resource Locators (URLs) and the manner by which near-English names are converted into IP addresses. We will also examine the growth in the use of the Internet and how this growth results in the need for techniques that conserve the use of IP addresses as well as the creation of a *next generation* (ng) IP address structure. Because this is an introductory chapter, there will be a sneak preview of the contents of succeeding chapters in this book. You may use this preview information either as is or in conjunction with the index to locate information of immediate interest.

While this book is structured on a chapter-by-chapter basis to be read in sequence, it was also recognized that in a hectic work environment many professions need to immediately access information of interest. Recognizing this, wherever possible each chapter was written to be as independent as possible of preceding and succeeding chapters. Thus, while it is suggested that students and professions that require a strong understanding of IP addressing read each chapter in sequence, it is possible to turn to a chapter to obtain specific information concerning an IP addressing topic of interest. With this said, it is now time to grab a Coke or Pepsi, relax, and follow me into the wonderful world of IP addressing.

Rationale

To obtain an appreciation for the need for IP addressing, we will turn our attention to several topics in this section. First, how routers operate will be

1

described and discussed in general terms to obtain an appreciation for the role of IP addresses in the routing process. Once this is accomplished, we will move on to a term referred to as address space, using a few simple computations to understand the reason why the current version of the IP has a finite number of addresses available for use and why different techniques that involve IP addressing became necessary to conserve address space.

Continuing in this section, we will briefly examine how we can use near-English terms to reference documents located on different computers, such as the home page of a Web server. While this allows us to avoid the direct use of IP addresses in our daily operations, we will also note that we must configure devices to enable them to initiate a name to address translation process. Once we appreciate the need for having an applicable computer configuration, we will discuss the use of the current version of IP versus the next generation version, referred to as IPv6. In doing so we will note the development of several IP addressing applications that extend the life of the current version of IP and result in the necessity for network managers and administrators to become conversant with the operation and utilization of those applications.

Router Operations

In the evolution of the Internet, routers were first referred to as gateways because they were initially developed to provide a connection or gateway function from one network to another. Although the functionality of routers has significantly increased over the past two decades, their primary function continues as a gateway to route data between networks. In fact, the term *gateway* continues to be used in most computer operating systems. When you configure a workstation or server to operate as a participant on a TCP/IP network, more than likely you will be requested to enter the IP address of a gateway. This address tells your workstation or server where to forward packets that are destined to another network other than the network your device resides on.

Exhibit 1 illustrates the relationship between a computer located on a local area network (LAN) that needs to access information on a different network and a router or gateway that provides a connection between networks. In examining Exhibit 1 note that the computer on the LAN is configured with the address of the router or gateway. Thus, it knows the location where to transmit information destined for outside the confines of the local network. That location is the address of the router, which in this example routes the data off the local network and into the Internet on its path toward its ultimate destination.

Use of IP Addresses

In examining Exhibit 1 it is important to note that all routing, whether on the Internet or on a private TCP/IP network, occurs using IP addresses. Thus, in addition to configuring equipment with IP addresses, different networking

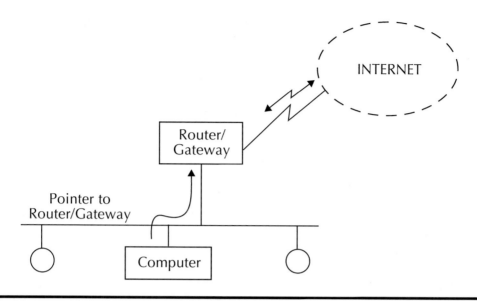

Exhibit 1.   Router or Gateway Transferring Packets Off the Local Network

devices use IP addresses as a decision criteria during the routing process. Now that we have a general appreciation for the role of IP addressing in the configuration of equipment and the routing of data, consider the fact that the number of IP addresses is limited by what is referred to as IP address space.

Address Space Availability

We can obtain additional appreciation for the need to understand different aspects of IP addressing by turning our attention to the address space supported by the current version of the IP, IPv4. IPv4 uses 32-bit addresses. A 32-bit destination address is used to identify the recipient of information, while a 32-bit source address is used to identify the originator of information being transmitted. On a formal basis, IP information is transmitted in units referred to as datagrams. However, in this chapter we will simply call information either by that term or refer to pieces of information as packets, leaving it to succeeding chapters to investigate the manner by which data is actually transported on a TCP/IP network.

When the TCP/IP protocol suite was developed during the 1960s, it was primarily viewed as a mechanism to interconnect research laboratories and universities. It is highly doubtful that anyone involved in the development of the TCP/IP protocol suite during the 1960s envisioned that its use would expand to the point that by the beginning of the new millennium, it would serve as a mechanism to interconnect hundreds of millions of persons on every continent on the globe. While the developers of the TCP/IP protocol suite may not have had a sharply focused crystal ball, they did employ an open architecture that enabled new applications to overcome some of the constraints of the protocol suite. In this section we will perform a few computations to denote some constraints and then discuss methods that were developed to overcome those constraints. In doing so we will become aware of additional reasons why it is necessary for many network managers and LAN administrators to literally move beyond the basics of IP addressing and obtain knowledge of tools and techniques that facilitate the economical use of IP addresses.

The use of a 32-bit address structure under IPv4 appeared to provide a significant number of distinct addresses during the 1960s that would enable every device that required a connection to the Internet to be connected. If we remember the binary number system, which will be reviewed in Chapter 3, each bit position represents a power of two. Exhibit 2 indicates the value of different powers of 2 from 0 to 31 which represents 32-bit positions used by IPv4 addressing. This table also indicates the bit position of each bit in a 32-bit address. Because the 32 bits in an IPv4 address are numbered from 0 to 31 to correspond with powers of 2, the position is always 1 above the power used to compute the value of the position.

In examining the entries in Exhibit 2, note that the value of position n represents the sum of all possible values through the power of $n - 1$. For example, when n is 9, the value of position 9 is 256. This represents the range of possible values that can be assigned to an 8-bit byte since its value can range from 0 to 255. Thus, an easy method to determine the theoretical maximum number of distinct addresses that could be supported under IPv4 would be to simply double the value for position 32 shown in Exhibit 2. Doing so, we obtain a value of 4,294,967,296.

While a value that exceeds four billion was certainly a large value in the 1960s, during the past 40 years the proliferation of personal computers, fax machines, PDAs, mobile phones, and other electronic devices that require access to the Internet has severely taxed the address space of IPv4. Recognizing the potential shortage of IPv4 addresses resulted in two significant areas of activity. One area was focused on developing a new IP protocol that would significantly expand IP address space. A second area of activity involved managing scarce IPv4 addresses more efficiently. This resulted in the use of the Dynamic Host Configuration Protocol (DHCP) and Network Address Translation (NAT) as techniques to more efficiently govern the use of IPv4 addresses. Thus, for some readers, knowledge of DHCP and NAT will provide the ability to understand how to better manage scarce IPv4 addresses and will be covered in this book. Now that we have a general appreciation for the reason why many readers will need to know information beyond basic IPv4 addressing to include applications that manage addresses, let us turn our attention to how the Internet works. In doing so we will briefly examine URLs, host names, IP addresses, and the role of the DNS to further reinforce the need for knowledge about a variety of IP addressing techniques.

Internet Operations

If our use of the Internet until now has been one of surfing the Web, our primary mechanism for addressing has been the use of URLs. A URL is an

Exhibit 2.   Powers of 2

Power	Position	Value
0	1	1
1	2	2
2	3	4
3	4	8
4	5	16
5	6	32
6	7	64
7	8	128
8	9	256
9	10	512
10	11	1024
11	12	2048
12	13	4096
13	14	8192
14	15	16384
15	16	32768
16	17	65536
17	18	131072
18	19	262144
19	20	524288
20	21	1048576
21	22	2097152
22	23	4194304
23	24	8388608
24	25	16777216
25	26	33554432
26	27	67108864
27	28	134217728
28	29	268435456
29	30	536870912
30	31	1073741824
31	32	2147483648

acronym for Uniform Resource Locator and represents an extension of a file name concept to point to information that can reside on any computer on a network. In surfing the Web, we probably became familiar with the entry of such addresses as:

http://www.yahoo.com
http://www.lycos.com
http://www.whitehouse.gov

Depending on the browser we are using, many times it is possible to simply enter a portion of an address, such as whitehouse.gov, and have the home page of the White House server be displayed on our browser. However, even though the ability to use abbreviations expedites productivity, let us take a step back and examine what the preceding addresses in the form of URLs really mean.

The term *http* stands for Hypertext Transport Protocol. HTTP represents the protocol used to transport information between Web browsers and Web servers. As part of the format of a URL, the protocol is followed by a colon (:) and two forward slashes (//). Thus, this explains the use of http:// in the preceding URL.

The information to the right of the pair of forward slashes represents what is referred to as the fully qualified domain name (FQDN) of the destination computer. That name can be considered to represent a position in an inverted tree structured directory where the suffix (.gov) indicates the address is part of the government domain space. Prefixing the suffix of .gov is .whitehouse, which, when combined with .gov, indicates that the address we seek is the White House, which is located in the government domain space. Because the White House could have many types of computers, a mechanism is required to identify individual computers. That mechanism is obtained by adding a host computer name to the previous information. Thus, the fully qualified domain name becomes www.whitehouse.gov.

When you enter the URL into your browser, several operations will occur transparently. First, your computer will examine an area of memory referred to as cache memory to determine if the IP address associated with the URL was previously learned. If so, your browser will use the IP address to create a packet that will request the display of the home page of the White House Web server. If the IP address was not previously learned, it must be determined because all routing is based on IP addresses and not URLs.

When your TCP/IP protocol was installed, another IP address was configured that enables the resolution of host names contained in URLs to IP addresses. That IP address is the location of a Domain Name Server (DNS) that provides a translation capability between host names and IP addresses. The DNS address in effect tells your computer where to go to determine the IP address required to form a packet for routing. Later in this book we will examine the operation of the DNS; however, for now we only need to know that this parallel system enables us to use near-English addresses that are obviously much easier to remember than the composition of a 32-bit destination address.

While many aspects of the DNS are transparent to most users, the ability to use this system requires computers to know the location of at least one DNS server. Thus, once again it is important to understand IP addressing to include the configuration of appropriate addresses in various configuration screens.

IPv4 versus IPv6

The protocol we currently use on the Internet is referred to as IPv4. This protocol uses a 32-bit address, which, as previously noted, provides slightly more than four billion unique addresses. Recognizing that growth in the use of the Internet could result in the eventual depletion of IPv4 address space, work commenced on a next generation Internet protocol. Originally referred to as IPng, this protocol was later labeled IPv6.

Under IPv6, an extended addressing scheme is used which results in 128 bits being available for destination and source addresses. Because each bit position doubles the number of addresses, it is probably possible for IPv6 to allow every grain of sand in the Sahara desert to obtain an individual IP address.

Although work was begun on IPv6 many years ago in recognition of the need to compensate for the rapid depletion of IPv4 address space, the use of DHCP, NAT, subnetting, and other address economizing techniques has literally postponed the day of reckoning. Thus, although IPv6 will be covered in this book, also described and discussed will be address economizing techniques that may alleviate the necessity to migrate to a new version of IP for many organizations. However, for other organizations, knowledge of IPv6 may be needed sooner rather than later. Thus, once again it becomes obvious that for many persons it is important to obtain a broad base of knowledge concerning IP addressing, including both applications and protocols.

Book Preview

To conclude this chapter, an overview of material to be presented in succeeding chapters will be provided. This information can be used either as is or in conjunction with the index to locate information of immediate need. It is recommended that readers go through the material presented in this book in the sequence presented. However, it is also recognized that often specific information is required and there is not time to sequence through an entire book to obtain information about a specific topic. Thus, where possible, material in each of the chapters was written to be as independent as possible from preceding and succeeding chapters.

The TCP/IP Protocol Suite

Any book on IP addressing needs to indicate the relationship of the various components of the TCP/IP protocol suite to IP addresses. Chapter 2 will focus on this topic. In this chapter, we will examine the layered architecture used by the TCP/IP protocol suite. We will examine both upper and lower layer protocols, the formation of datagrams, and the manner by which different types of addresses are resolved. Although we will note the location and use of IP addresses, a detailed examination of IPv4 addresses and the role of the DNS will be deferred until Chapter 3.

IPv4 Addressing Basics and the DNS

Because IPv4 represents the current protocol and addressing method used on the majority of the Internet, we will commence our examination of the wonderful world of IP addressing with this topic and a related topic. In Chapter 3 we will examine the basic structure of IPv4 addresses, how work-stations are configured, and the role of the DNS.

Because readers more than likely have diverse backgrounds, we will initiate a short review of binary mathematics and note the use of dotted decimal notation in forming an IPv4 address.

The Role of Special IP Addresses

Continuing our probe into IP addressing, we will review the role of special IP addresses in Chapter 4. Chapter 4 will examine the use of private IP addresses referred to as RFC 1918 addresses, loopback addresses, and other addresses not commonly used on private IP networks as well as on networks connected to the Internet. The information presented in Chapter 4 will then be used to discuss a term referred to as address spoofing by which hackers attempt to hide their true IP address when they attack a host. Because it is desirable to learn how to block such addresses, we will also turn our attention to how we can configure a router to provide protection against datagrams containing commonly spoofed IP addresses from flowing onto our organization's network from the Internet.

Subnetting

Subnetting was developed as a mechanism to conserve IP addresses as well as to reduce router table entries. Chapter 5 will examine how we can design a network consisting of several subnets to include developing an appropriate addressing scheme. In addition, we will also note the role of the subnet mask and the manner by which participants on a network must be configured to correctly operate on a subnet.

CIDR and Multicasting

Chapter 6 will focus on two topics gaining widespread interest among many organizations — classless inter-domain routing (CIDR) and multicasting. CIDR provides organizations with the ability to make more efficient use of IPv4 addresses, while multicasting provides organizations with the ability to more efficiently transmit information destined to multiple users.

NAT and Naming Services

Chapter 7 will focus on two topics that resulted in the ability to significantly conserve on the use of IPv4 addresses. The topics we will examine include Network Address Translation (NAT), which can permit one IP address to be used to support thousands to tens of thousands of computers accessing the Internet, and two IP naming services that enable IP addresses to be dynamically reused. By reading this chapter we will note why we can paraphrase Mark Twain and say that the death of IPv4 is greatly exaggerated.

Working With IPv6

Although the death of Mark Twain might be greatly exaggerated, eventually things do come to an end. Eventually the last IPv4 address will be allocated and the Internet community will have to migrate to IPv6, which is the focus of Chapter 8. Chapter 8 will examine the structure of IPv6, its enhanced addressing capability, and the improvements provided by the protocol which extend beyond an increased addressing capability.

Network Utility Tools

In concluding this book we will turn our attention to the use of a core set of network utility tools that facilitate the testing and troubleshooting of TCP/IP networks. Chapter 9 will examine the old workhorses known as Ping and Traceroute as well as the more modern Pingpath introduced under Windows 2000. Therefore, if you are still sipping a Coke or Pepsi, relax a moment and follow me to the next chapter to obtain an appreciation of the TCP/IP protocol suite.

Chapter 2

The TCP/IP Protocol Suite

The primary purpose of Chapter 2 is to obtain an appreciation for the general composition of the TCP/IP protocol suite. To accomplish this task, we will first examine the International Standards Organization (ISO) Open Systems Interconnection (OSI) Reference Model. Once this is accomplished, we will examine the Internet Protocol (IP), which represents a network layer protocol. This will be followed by moving up the protocol stack and examining operations at the transport layer.

The OSI Reference Model

Although the TCP/IP protocol suite predated the OSI Reference Model, we can obtain a better appreciation for the functioning of the TCP/IP protocol suite by examining the layering concept associated with communications defined by that model. During the 1970s, approximately a dozen years after the development of several popular communications protocols to include TCP/IP, the International Standards Organization (ISO) established a framework for standardizing communications systems. This framework was called the Open System Interconnection (OSI) Reference Model and it defines an architecture in which communications functions are divided into seven distinct layers, with specific functions becoming the responsibility of a particular layer.

Exhibit 1 illustrates the seven layers of the OSI Reference Model. Note that each layer, with the exception of the lowest, covers a lower layer, effectively isolating it from higher layer functions. Layer isolation is an important aspect of the OSI Reference Model because it allows the given characteristics of one layer to change without affecting the remainder of the model, provided that support services remain the same. This is possible because well-known interface points in a layered model enable one layer to communicate with another, even though one or both may change. In addition, the layering process permits

Layer 7	Application
Layer 6	Presentation
Layer 5	Session
Layer 4	Transport
Layer 3	Network
Layer 2	Data Link
Layer 1	Physical

Exhibit 1.   The OSI Open System Interconnection Reference Model

end users to mix and match OSI or other layered protocol-conforming communications products to tailor their communications system to satisfy a particular networking requirement. Thus, the OSI Reference Model, as well as protocol suites that employ a layered architecture, provides the potential to directly interconnect networks based on the use of different vendor products. This architecture, which is referred to as an *open architecture* when its specifications are licensed or placed in the public domain, can be of substantial benefit to both users and vendors. For users, an open architecture removes them from dependence on a particular vendor and may also prove to be economically advantageous because it fosters competition.

For vendors, the ability to easily interconnect their products with products produced by other vendors opens up a wider market. Now that we have an appreciation for the value of a layered architecture, let us turn our attention to the functions of the seven layers of the OSI Reference Model.

OSI Reference Model Layers

As previously noted, the OSI Reference Model consists of seven layers, with specific functions occurring at each layer. In this section we will discuss the functions performed at each layer in the OSI Reference Model. We will use this information in the next section in this chapter to better understand the components of the TCP/IP protocol suite.

Layer 1: The Physical Layer

The physical layer represents the lowest layer in the OSI Reference Model. Because the physical layer involves the connection of a communications system to a communications media, the physical layer is responsible for specifying the electrical and physical connection between communications devices that connect to the different types of media. At this layer, cable connections and the electrical rules necessary to transfer data between devices are specified. Examples of physical layer standards include RS-232, V.24, and the V.35 interface.

Layer 2: The Data Link Layer

The second layer in the OSI Reference Model is the data link layer. This layer is responsible for defining the manner by which a device gains access to the medium specified in the physical layer. In addition, the data link layer is also responsible for defining data formats to include the entity by which information is transported, error control procedures, and other link control procedures.

Most trade literature and other publications reference the entity by which information is transported at the data link layer as a frame. Depending on the protocol used, the frame will have a certain header composition with fields that normally indicate the destination address and the originator of the frame through the use of a source address. In addition, frames will have a trailer with a cyclic redundancy check (CRC) field that indicates the value of an error checking mechanism algorithm performed by the originator on the contents of the frame. The receiver will apply the same algorithm to an inbound frame and compare the locally generated CRC to the CRC in the trailer. If the two match, the frame is considered to be received without error, while a mismatch indicates a transmission error occurred, and the receiver will then request the originator to retransmit the frame. Examples for common Layer 2 protocols include such LAN protocols as Ethernet and Token-Ring as well as such WAN protocols as High-Level Data Link Control (HDLC).

The original development of the OSI Reference Model targeted wide area networking. This resulted in its applicability to LANs requiring a degree of modification. The Institute of Electrical and Electronic Engineers (IEEE), which is responsible for developing LAN standards, subdivided the data link layer into two sublayers: (1) Logical Link Control (LLC) and (2) Media Access Control (MAC). The LLC layer is responsible for generating and interpreting commands that control the flow of data and performing recovery operations in the event errors are detected. In comparison, the MAC layer is responsible for providing access to the local area network, which enables a station on the network to transmit information. The subdivision of the data link layer allows a common LLC layer to be used regardless of differences in the method of network access. Thus, a common LLC is used for both Ethernet and Token-Ring although their access methods are dissimilar.

Layer 3: The Network Layer

Moving up the OSI Reference Model, the third layer is the network layer. This layer is responsible for arranging a logical connection between a source and destination on the network to include the selection and management of a route for the flow of information between source and destination based on the available paths within a network.

Services or functions provided at the network layer are associated with the movement of data through a network to include addressing, routing, switching, sequencing, and flow control procedures. At the network layer, units of information are placed into packets that have a header and trailer similar to frames at the data link layer. Thus, the network layer packet will contain addressing information as well as a field that facilitates error detection and correction.

In a complex network, the source and destination may not be directly connected by a single path. Instead, a path may be required to be established through the network that consists of several subpaths. Thus, the routing of packets through the network as well as the mechanism in the form of routing protocols that provide information about available paths are important features of this layer.

Several protocols are standardized for Layer 3 to include the International Telecommunications Union Telecommunications body (ITU-T) X.25 packet switching protocol and the ITU-TX75 gateway protocol. X.25 governs the flow of information through packet network, whereas X.75 governs the flow of information between packet networks. When we examine the TCP/IP protocol suite in succeeding sections in this chapter, note that the IP represents the network layer protocol used in the TCP/IP protocol suite.

Layer 4: The Transport Layer

Continuing our tour of the OSI Reference Model, the transport layer is responsible for governing the transfer of information after a route has been established through the network by the network layer protocol. There are two general types of transport layer protocols: (1) connection-oriented and (2) connectionless. A connection-oriented protocol first requires the establishment of a connection prior to data transfer occurring. This type of transport layer protocol performs error control, sequence checking, and other end-to-end data reliability functions. A second type or category of transport layer protocol operates as a connectionless, best-effort protocol. This type of protocol depends on higher layers in the protocol suite for error detection and correction. As we examine the TCP/IP protocol suite, we will note that TCP represents a Layer 4 connection-oriented protocol while UDP represents a connectionless Layer 4 protocol.

Layer 5: The Session Layer

The fifth layer in the OSI Reference Model is the session layer. This layer is responsible for providing a set of rules that govern the establishment and termination of data streams flowing between nodes in a network. The services that the session layer can provide include establishing and terminating node connections, message flow control, dialogue control, and end-to-end data control. In the TCP/IP protocol suite, Layers 5 through 7 are grouped together as an application layer.

Layer 6: The Presentation Layer

The sixth layer of the OSI Reference Model is the presentation layer. This layer is concerned with the conversion of transmitted data into a display format appropriate for a receiving device. This conversion can include data codes as well as display placement. Other functions performed at the presentation layer can include data compression and decompression and data encryption and decryption.

Layer 7: The Application Layer

The top layer of the OSI Reference Model is the application layer. This layer functions as a window through which the application gains access to all of the services provided by the model. Examples of functions performed at the application layer include electronic mail, file transfers, resource sharing, and data base access. Although the first four layers of the OSI Reference Model are fairly well defined, the top three layers can vary considerably between networks. As previously mentioned, the TCP/IP protocol suite, which is a layered protocol that predates the ISO Reference Model, combines Layers 5 through 7 into one application layer.

Data Flow

The design of an OSI Reference Model compatible network is such that a series of headers are opened to each data unit as packets are transmitted and delivered by frames. At the receiver, the headers are removed as a data unit flows up the protocol suite, until the *headerless* data unit is identical to the transmitted data unit. In the next section of this chapter, we will examine the flow of data in a TCP/IP network that follows the previously described OSI Reference Model data flow.

The OSI Reference Model never lived up to its intended goal, with ISO protocols achieving a less than anticipated level of utilization. The concepts of the model made persons aware of the benefits that could be obtained by a layered open architecture as well as the functions that would be performed by different layers of the model. Thus, the ISO succeeded in making networking personnel aware of the benefits that could be derived from a layered open architecture and more than likely contributed to the success of the acceptance of the TCP/IP protocol suite. We will now turn our attention to the TCP/IP protocol suite.

Overview of the TCP/IP Protocol Suite

The Transmission Control Protocol/Internet Protocol (TCP/IP) actually represents two distinct protocols within the TCP/IP protocol suite. Due to the popularity of those protocols, and the fact that a majority of traffic is transferred using those protocols, the members of the protocol suite include TCP and IP and are collectively referred to as TCP/IP.

   Exhibit 2 provides a general comparison of the structure of the TCP/IP protocol suite to the OSI Reference Model. The term *general comparison* is used because the protocol suite consists of hundreds of applications, of which only a handful is shown. Another reason that Exhibit 2 is a general comparison results from the fact that the TCP/IP protocol suite actually begins above the data link layer. Although the physical and data link layers are not part of the TCP/IP protocol suite, they are shown in Exhibit 2 to provide a frame of reference to the OSI Reference Model as well as to facilitate an explanation of the role of two special protocols within the TCP/IP protocol suite.

OSI Reference   TCP/IP Protocol Suite
Model Layer

Application							Other Applications
Presentation	FTP	HTTP	Telnet	SMTP	DNS	SNMP	Other Applications
Session							Other Applications
Transport	TCP				UDP		
Network			ICMP				
Network			IP				
Network			ARP				
Data Link	Ethernet		Token Ring	X.25		Frame Relay	Other
Physical	Physical Layer						

Exhibit 2.   Comparing the TCP/IP Protocol Suite to the OSI Reference Model

The Network Layer

The network layer of the TCP/IP protocol stack primarily consists of the IP. The IP protocol includes an addressing scheme that identifies the source and destination address of the packet being transported. In TCP/IP terminology, the unit of data being transmitted at the network layer is referred to as a datagram. Also included in what can be considered to represent the network protocols are the address resolution protocol (ARP) and the Internet Control Message Protocol (ICMP).

IP

The IP provides the addressing capability that allows datagrams to be routed between networks. The current version of IP is IPv4, under which IP addresses consist of 32 bits. There are currently five classes of IP addresses, referred to as Class A through Class E, with Classes A, B, and C having their 32 bits subdivided into a network portion and a host portion. The network portion of the address defines the network where a particular host resides, while the host portion of the address identifies a unique host on the network. Later in this chapter we will examine the IP in detail. However, we will defer until Chapter 3 a detailed examination of IP addressing.

ARP

One of the more significant differences between the data link layer and the network layer is the method of addressing used at each layer. At the data link layer, such LANs as Ethernet and Token-Ring networks use 48-bit MAC addresses. In comparison, TCP/IP currently uses 32-bit addresses under the current version of the IPD protocol, and the next generation of the IP protocol, IPv6, uses a 128-bit address. Thus, the delivery of a packet or datagram flowing at the network layer to a station on the LAN requires an address conversion. That address conversion is performed by the Address Resolution Protocol whose operation is covered in detail later in this chapter.

ICMP

The Internet Control Message Protocol (ICMP), as its name implies, represents a protocol used to convey control messages. Such messages range in scope from routers responding to a request that cannot be honored with a *destination unreachable* message to the requester, to messages that convey diagnostic tests and responses. An example of the latter is the echo-request/echo-response pair of ICMP datagrams that is more popularly referred to collectively as Ping.

ICMP messages are conveyed with the prefix of an IP headed to the message. Thus, we can consider ICMP to represent a Layer 3 protocol in the TCP/IP protocol suite. We will examine the structure of ICMP messages as well as the use of certain messages later in this chapter when we examine the network layer of the TCP/IP protocol suite detail.

The Transport Layer

As indicated in Exhibit 2, there are two transport layer protocols supported by the TCP/IP protocol suite: (1) the Transmission Control Protocol (TCP) and (2) the User Datagram Protocol (UDP).

TCP

TCP (Transmission Control Protocol) is an error-free, connection-oriented protocol. This means that prior to data being transmitted by TCP, the protocol requires the establishment of a path between source and destination as well as an acknowledgment that the receiver is ready to receive information. Once the flow of data commences, each unit, which is referred to as a TCP segment, is checked for errors at the receiver. If an error is detected through a checksum process, the receiver will request the originator to retransmit the segment. Thus, TCP represents an error-free, connection-oriented protocol.

The advantages associated with the use of TCP as a transport protocol relate to its error-free, connection-oriented functionality. For the transmission of relatively large quantities of data or important information, it makes sense to use this transport layer protocol. The connection-oriented feature of the protocol means that it will require a period of time for the source and destination to exchange handshake information. In addition, the error-free capability of the protocol may be redundant if the higher layer in the protocol suite also performs error checking. Recognizing the previously mentioned problems, the developers of the TCP/IP protocol suite added a second transport layer protocol referred to as UDP.

UDP

The User Datagram Protocol (UDP) is a connectionless, best-effort, non-error checking transport protocol. UDP was developed in recognition of the fact that some applications may require small pieces of information to be transferred, and the use of a connection-oriented protocol would result in a significant overhead to the transfer of data. Because a higher layer in the protocol suite could perform error checking, error detection and correction could also be eliminated from UDP. Because UDP transmits a piece of information referred to as a UDP datagram without first establishing a connection to the receiver, the protocol is also referred to as a *best-effort* protocol. To ensure that a series of UDP datagrams are not transmitted into a black hole if a receiver is not available, the higher layer in the protocol suite using UDP as a transport protocol will wait for an acknowledgment. If one is not received within a predefined period of time, the application can decide whether to retransmit or cancel the session.

In examining Exhibit 2 you will note that certain applications use TCP as their transport protocol while other applications use UDP. In general, applications that require data integrity, such as remote terminal transmission (Telnet), file transfer (FTP), and electronic mail, use TCP as their transport protocol. In comparison, applications that transmit relatively short packets, such as the Domain Name Service (DNS) and the Simple Network Management Protocol (SNMP) that is used to perform network management operations, use UDP.

One relatively new TCP/IP application takes advantage of both the TCP and UDP transport protocols. That application is voice over IP (VoIP). VoIP commonly uses TCP to set up a call and convey signaling information to the distant party. Because real time voice cannot be delayed by retransmission if an error in a packet is detected, there is no need to perform error detection. Thus, digitized voice samples are commonly transmitted using UDP once a session is established using TCP.

Application Layer

As previously noted, the development of the TCP/IP protocol suite predated the development of the ISO's OSI Reference Model. At the time the TCP/IP protocol suite was developed, functions above the transport layer were combined into one entity that represented an application. Thus, the TCP/IP protocol suite does not include separate session and presentation layers. Now that we have an appreciation for the manner by which the TCP/IP protocol stack can be compared and contrasted to the OSI Reference Model, we will conclude this section by examining the flow of data within a TCP/IP network.

Data Flow

Data flow within a TCP/IP network commences at the application layer where data is provided to an applicable transport layer protocol-TCP or UDP. At Layer 4, the transport layer opens either a TCP or a UDP header to the application data depending on the transport protocol used by the application layer.

The transport layer protocol uses a port number to distinguish the type of application data being transported. Through the use of port numbers, it becomes possible to distinguish one application from another that flows between a common source and destination.

For lay persons not familiar with TCP or IP, this explains how a common hardware platform, such as a Windows NT server, can support both Web and FTP services. That is, although the server has a common IP address contained in an IP header, the port number in the TCP or UDP header indicates the application.

Application data flowing onto a network is first formed into a TCP segment or UDP datagram. The resulting UDP datagram or TCP segment is then passed to the network layer where an IP header is opened. The IP header contains network addressing information that is used by routers to route datagrams through a network.

When an IP datagram reaches a LAN, the difference between the network layer and LAN address is first resolved through ARP. Once this is accomplished, the IP datagram is placed into a LAN frame using an appropriate MAC address in the LAN header. Exhibit 3 illustrates the data flow within a TCP/IP network for deliver to a station on a LAN.

The TCP/IP protocol suite represents a methodically considered and developed collection of protocols and applications. As we will note in subsequent sections of this chapter, it is a very flexible open architecture that allows new applications and protocols to be developed.

The Internet Protocol and Related Protocols

In this section attention will focus on what this author commonly refers to as the Network Layer Troika of the TCP/IP protocol suite, IP, ARP, and ICMP. In examining the IP, we will pay particular attention to the structure of the IP header and its fields, which are examined by routers as a mechanism for making forwarding decisions. Although we will defer a detailed discussion of IP addressing until Chapter 4, we will discuss several little known areas of the IP protocol. Having knowledge about these areas can provide us with network design and operation flexibility. One such topic is the assignment of multiple network addresses to an interface.

Initial focus in this section is on the IP protocol to include its use for routing datagrams across a network and between interconnected networks. In doing so, we will examine in detail the composition of the IP header and the use of different fields in the header. Once this is accomplished, attention will be on the role and operation of the Address Resolution Protocol (ARP) to include examining the rationale for a little-known ARP technique that can considerably facilitate the operation of delay sensitive transmissions, such as voice over IP. This section will conclude with Internet Control Message Protocol (ICMP).

The Internet Protocol

The IP represents the network layer of the TCP/IP protocol suite. IP was developed as a mechanism to interconnect packet switched TCP/IP based networks to form an internet. Here the term *internet* with a lower case *i* is used to represent the connection of two or more TCP/IP based networks.

Datagrams and Segments

The IP transmits blocks of data referred to as datagrams. As indicated in Exhibit 3, IP receives upper layer protocol data containing either a TCP or UDP header, referred to as a TCP segment or UDP datagram. The prefix of an IP header to the TCP segment or UDP datagram results in the formation of an IP datagram. This datagram contains a destination IP address that is used for routing purposes.

Datagrams and Datagram Transmissions

To alleviate potential confusion between datagrams and an obsolete transmission method referred to as datagram transmission, a few words are in order. When the ARPAnet evolved, two methods of packet transmission were experimented with. One method was referred to as datagram transmission and avoided the use of routers to perform table lookups. Under datagram transmission, each node in a network transmits a received datagram onto all ports other than the port the datagram was received on. While this technique avoids the need for routing table lookup operations, it can result in duplicate datagrams being received at certain points within a network. This results in the necessity to develop software to discard duplicate datagrams, adding an additional level of complexity to networking. Thus, datagram transmission was soon discarded in favor of the creation of virtual circuits that represent a temporary path established between source and destination. In the remainder of this book, when I refer to datagram transmission, I will actually be referencing the transmission of datagrams via a virtual circuit created between source and destination.

Routing

The actual routing of an IP datagram occurs on a best-effort or connectionless delivery mechanism. This is because IP by itself does not establish a session between the source and destination before it transports datagrams. When IP transports a TCP segment, the TCP header results in a connection-oriented session between two Layer 4 nodes transported by IP as a Layer 3 network protocol.

The importance of IP can be noted by the fact that routing between networks is based on IP addresses. As we will note later in this chapter, the device that routes data between different IP addressed networks is known as a router. Because it would be extremely difficult, if not impossible, to statically configure every router in a large network to know the route to other routers and networks connected to different routers, routing protocols are indispensable to the operation of a dynamic series of interconnected IP networks. Because the best way to obtain an appreciation for the operation of the IP is through an examination of the fields in its header, let us do so.

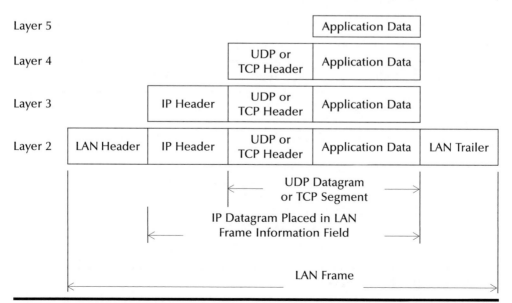

Exhibit 3. Data Flow within a TCP/IP Network for Delivery to a Station on a LAN

The IP Header

The current version of the IP is version 4, resulting in IP commonly referred to as IPv4. The next generation of the IP is IPv6. This section will focus attention on IPv4, leaving until Chapter 8 examination of IPv6.

Exhibit 4 illustrates the fields contained in the IPv4 header. In examining the IPv4 header illustrated in Exhibit 4, note that the header consists of a minimum of 20 bytes of data, with the width of each field shown with respect to a 32-bit (4-byte) word.

Bytes versus Octets

In this book, the term *byte* will be used to reference a sequence of 8 bits used in a common manner. During the development of the TCP/IP protocol suite and continuing today, most standards documents use the term *octet* to reference a collection of 8 bits. The use of the term *octet* is due to differences in the composition of a byte during the 1960s.

During the early development of computer systems, differences in computer architecture resulted in the use of groupings of 5 to 10 bits to represent a computer byte. Thus, the term byte at that time was ambiguous, and standard-setting bodies decided to use the term *octet* to reference a grouping of 8 bits. Because all modern computers use 8-bit bytes, the term *byte* is no longer ambiguous. Thus, we will use the term *byte* throughout this book.

To appreciate the operation of IP, let us examine the functions of the fields in the header. As we do so, we will, when appropriate, discuss the relation of certain fields to routing and security, topics that will be discussed in detail in later chapters of this book.

0	4	8	16	31
Vers	Hlen	Service Type	Total Length	
Identification			Flag	Fragment Offset
Time to Live		Protocol	Header Checksum	
Source IP Address				
Destination IP Address				
Options + Padding				

Exhibit 4.   The IPv4 Header

Vers Field

The Vers field is 4 bits in length and is used to identify the version of the IP protocol used to create an IP datagram. The current version of the IP protocol is v4, with the next generation of the IP protocol assigned version number 6.

The 4 bits in the Vers field support 16 version numbers. Under RFC 1700, a listing of Internet version numbers can be obtained and a summary of that listing is in Exhibit 5. In examining Exhibit 5, the reason the next generation IP is IPv6 instead of IPv5 is related to the fact that version 5 was previously assigned to an experimental protocol referred to as the Streams 2 Protocol.

Hlen Field

The length of the IP header can vary due to its ability to support options. To allow a receiving device to correctly interpret the contents of the header from the rest of an IP datagram requires the receiving device to know where the header ends. This function is performed by the Hlen field whose value indicates the length of the header.

The Hlen field is 4 bits in length. If we examine Exhibit 4 we will note that the IP header consists of 20 bytes of fixed information followed by options. Because it is not possible to use a 4-bit field to directly indicate the length of a header equal to or exceeding 320 bytes, the value in this field represents the number of 32-bit words in the header. For example, the shortest IP header is 20 bytes, which represent 160 bits. When divided by 32 bits, this results in a value of 160/32 or 5, which is the value set into the Hlen field when the IP header contains 20 bytes and no options.

Exhibit 5. Assigned Internet Version Numbers

Numbers	Assignment
0	Reserved
1–3	Unassigned
4	IP
5	Streams
6	IPv6
7	TP/IX
8	P Internet Protocol (PIP)
9	TUBA
10–14	Unassigned
15	Reserved

Service Type Field

The Service Type field is an 8-bit field that is commonly referred to as a Type of Service (TOS) field.

The initial development of the IP protocol assumed that applications would use this field to indicate the type of routing path they would like. Routers along the path of a datagram would examine the contents of the Service Type byte and attempt to comply with the setting in this field.

Exhibit 6 illustrates the format of the Service Type field. This field consists of two subfields: (1) type of service and (2) precedence. The type of service subfield consists of bit positions that, when set, indicate how a datagram should be handled. The three bits in the precedence field allow the transmitting station to indicate to the IP layer the priority for sending a datagram. A value of 000 indicates a normal precedence, while a value of 111 indicates the highest level of precedence and is normally used for network control.

The value in the Precedence field is combined with a setting in the Type of Service field to indicate how a datagram should be processed. As indicated in the lower portion of Exhibit 6, there are six settings defined for the Type of Service field. To understand how this field would be used, assume an application was transmitting digitized voice that requires minimal routing delays due to the effect of latency on the reconstruction of digitized voice. By setting the Type of Service field to a value of 1000, this would indicate to each router in the path between source and destination network that the datagram was delay sensitive and its processing by the router should minimize delay.

7	6	5	4	3	2	1	0
R	Type of Service				Precedence		

where:

       R represents reserved.

      Precedence provides eight levels, 0 to 7, with 0 normal and 7 the highest.

      Type of Service (TOS) indicates how the datagram is handled.

      0000   Default
      0001   Minimize monetary cost
      0010   Maximize reliability
      0100   Maximize throughput
      1000   Minimize delay
      1111   Maximize security

Exhibit 6.   The Service Type Field

In comparison, because routers are designed to discard packets under periods of congestion, an application where the ability of packets to reach their destination is of primary importance would set the TOS field to a value of 0010. This setting would denote to routers in the transmission path that the datagram requires maximum reliability. Thus, routers would select other packets for discard prior to discarding a packet with its TOS subfield set to a value of 0010.

Although the concept behind including a Service Type field was a good idea, from a practical standpoint it is rarely used. The reason for its lack of use is the need for routers supporting this field to construct and maintain multiple routing tables. While this is not a problem for small networks, the creation and support of multiple routing tables can significantly effect the level of performance of routers in a complex network, such as the Internet.

Although routers in most networks ignore the contents of the Service Type field, this field is now being used to map IP datagrams being transmitted over an ATM backbone. Because ATM includes a built-in Quality of Service (QOS) that, at the present time, cannot be obtained on an IP network, many organizations are transmitting a variety of data to include voice over IP over an ATM backbone, using the Service Type field as a mechanism to map different IP service requirements into applicable types of ATM service. A second emerging application for the Service Type field is to differentiate the requirements of different applications as they flow into an IP network. In this situation, the Service Type byte is renamed as the DiffServe (Differentiated Service) byte. The Internet Engineering Task Force is currently examining the potential use of the DiffServe byte as a mechanism to define an end-to-end quality of service capability through an IP network.

Total Length Field

The Total Length field indicates the total length of an IP datagram in bytes. This length indicates the length of the IP header to include options followed by a TCP or UDP header or another type of header that we will shortly discuss, as well as the data that follows that header. The total length field is 16 bits in length, resulting in an IP datagram having a maximum defined length of 2^{16} or 65536 bytes.

Identification and Fragment Offset Fields

Unlike some types of clothing where one size fits all, an IP datagram can range up to 65536 bytes in length. Because some networks only support a transport frame that can carry a small portion of the theoretical maximum length IP datagram, it can become necessary to fragment the datagram for transmission between networks. One example of this would be the routing of a datagram from a Token-Ring network to another Token-Ring network via an Ethernet network. Token-Ring networks that operate at 16 Mbps can transport approximately 18 Kbytes in their Information field. In comparison, an Ethernet frame has a maximum length Information field of 1500 bytes. This means that datagrams routed between Token-Ring networks via an Ethernet network must be subdivided or fragmented into a maximum length of 1500 bytes for Ethernet to be able to transport the data.

The default IP datagram length is referred to as the path MTU (Maximum Transmission Unit). The MTU is defined as the size of the largest packet that can be transmitted or received through a logical interface. When two Token-Ring networks are connected via an Ethernet network, the MTU would be 1500 bytes. Because it is important to commence transmission with the lowest common denominator packet size that can flow through different networks, and, if possible, adjust the packet size after the initial packet reaches its destination, IP datagrams use a default of 576 bytes when datagrams are transmitted remotely (off the current network).

Fragmentation is a most interesting function because it allows networks capable of transmitting larger packets to do so more efficiently. The reason efficiency increases is due to the fact that larger packets have proportionally less overhead. Unfortunately, the gain in packet efficiency is not without cost. First, although routers can fragment datagrams, they do not reassemble them, leaving it to the host to perform reassembly. This is because router CPU and memory requirements would considerably expand if they had to reassemble datagrams flowing to networks containing hundreds or thousands of hosts. Secondly, although fragmentation is a good idea for boosting transmission efficiency, a setting in the Flag field, which we will shortly cover, can be used to indicate a datagram should not be fragmented. Because many routers do not support fragmentation, many applications by default set the do not fragment flag bit and use a datagram length that, while perhaps not most efficient, ensures that a datagram can flow end-to-end as its length represents the lowest common denominator of the networks it will traverse.

When an IP datagram is fragmented, this situation results in the use of three fields in the IP header. Those fields are the Identification, Flag, and Fragment Offset fields.

The Identification field is 16 bytes in length and is used to indicate which datagram fragments belong together. A receiving device operation at the IP network layer uses the Identification field as well as the source IP address to determine which fragments belong together. To ensure fragments are put back together in their appropriate order requires a mechanism to distinguish one fragment from another. That mechanism is provided by the Fragment Offset field, which indicates the location where each fragment belongs in a complete message. The actual value in the Fragment Offset field is an integer that corresponds to a unit of 8 bytes that indicates the offset from the previous datagram. For example, if the first fragment were 512 bytes in length, the second fragment would have an offset value that indicates that this IP datagram commences at byte 513. By using the total length and Fragment Offset fields, a receiver can easily reconstruct a fragmented datagram.

Flag Field

The third field in the IP header directly associated with fragmentation is the Flag field. This field is 4 bytes in length, with 2 bits used to denote fragmentation information. The setting of one of those bits is used as a direct fragment control mechanism, since a value of *0* indicates the datagram can be fragmented, while a value of *1* indicates do not fragment the datagram. The second fragment bit is used to indicate fragmentation progress. When the second bit is set to a value of *0*, it indicates that the current fragment in a datagram is the last fragment. In comparison, a value of *1* in this bit position indicates that more fragments follow.

Time to Live Field

The Time to Live (TTL) field is 8 bits in length. The setting in this field is used to specify the maximum amount of time that a datagram can exist. It is used to prevent a misaddressed datagram from endlessly wandering the Internet or a private IP network, similar to the manner by which a famous American folk hero was noted in a song to wander the streets of Boston.

Because an exact time is difficult to measure, the value placed into the TTL field is actually a router hop count. That is, routers decrement the value of the TTL field by 1 as a datagram flows between networks. If the value of this field reaches zero, the router will discard the datagram and, depending on the configuration of the router, generate an ICMP message that informs the originator of the datagram that the TTL field expired and the datagram, in effect, was sent to the great bit bucket in the sky.

Many applications set the TTL field value to a default of 32, which should be more than sufficient to reach most destinations in a very complex network to include the Internet. In fact, one popular application referred to as Traceroute will issue a sequence of datagrams commencing with a value of 1 in the TTL field to obtain a sequence of router-generated ICMP messages that enable the path from source to destination to be noted. Chapter 9 will examine the operation of the Traceroute application and note how it can be used as a diagnostic tool.

Protocol Field

In Exhibit 3, we noted that an IP header prefixes the transport layer header to form an IP datagram. While TCP and UDP represent a large majority of Layer 4 protocols carried in an IP datagram, they are not the only protocols transported. In addition, even if they were, we would need a mechanism to distinguish one upper layer protocol from another that is carried in a datagram.

The method used to distinguish the upper layer protocol carried in an IP datagram is obtained through the use of a value in the Protocol field. For example, a value of decimal 6 is used to indicate that a TCP header follows the IP header, while a value of decimal 17 indicates that a UDP header follows the IP header in a datagram.

The Protocol field is 8 bits in length, permitting up to 256 protocols to be defined under IPv4. Exhibit 7 lists the current assignments of Internet protocol numbers. Note that although TCP and UDP by far represent the vast majority of TCP/IP traffic on the Internet and corporate intranets, other protocols can be transported and a large block of protocol numbers are currently unassigned. Also note that under IPv6, the Protocol field is named the Next Header field. Chapter 8 will examine IPv6 in detail.

Exhibit 7.   Assigned IP Numbers Field Values

Decimal	Keyword	Protocol
0	HOPOPT	IPv6 Hop-by-Hop Option
1	ICMP	Internet Control Message
2	IGMP	Internet Group Management
3	GGP	Gateway-to-Gateway
4	IP	IP in IP (encapsulation)
5	ST	Stream
6	TCP	Transmission Control Protocol
7	CBT	CBT
8	EGP	Exterior Gateway Protocol
9	IGP	Any private interior gateway (used by Cisco for their IGRP)
10	BBN-RCC-MON	BBN RCC Monitoring
11	NVP-II	Network Voice Protocol Version 2
12	PUP	PUP
13	ARGUS	ARGUS
14	EMCON	EMCON
15	XNET	Cross Net Debugger
16	CHAOS	Chaos
17	UDP	User Datagram
18	MUX	Multiplexing
19	DCN-MEAS	DCN Measurement Subsystems
20	HMP	Host Monitoring
21	PRM	Packet Radio Measurement
22	XNS-IDP	XEROX NS IDP
23	TRUNK-1	Trunk-1
24	TRUNK-2	Trunk-2
25	LEAF-1	Leaf-1
26	LEAF-2	Leaf-2
27	RDP	Reliable Data Protocol
28	IRTP	Internet Reliable Transaction
29	ISO-TP4	ISO Transport Protocol class 4
30	NETBLT	Bulk Data Transfer Protocol
31	MFE-NSP	MFE Network Services Protocol
32	MERIT-INP	MERIT Internodal Protocol
33	SEP	Sequential Exchange Protocol
34	3PC	Third Party Connect Protocol

Exhibit 7. (Continued)   Assigned IP Numbers Field Values

Decimal	Keyword	Protocol
35	IDPR	Inter-Domain Policy Routing Protocol
36	XTP	XTP
37	DDP	Datagram Delivery Protocol
38	IDPR-CMTP	IDPR Control Message Transport Protocol
39	TP++	TP++ Transport Protocol
40	IL	IL Transport Protocol
41	IPv6	Ipv6
42	SDRP	Source Demand Routing Protocol
43	IPv6-Route	Routing Header for IPv6
44	IPv6-Frag	Fragment Header for IPv6
45	IDRP	Inter-Domain Routing Protocol
46	RSVP	Reservation Protocol
47	GRE	General Routing Encapsulation
48	MHRP	Mobile Host routing Protocol
49	BNA	BNA
50	ESP	Encap security Payload for IPv6
51	AH	Authentication Header for IPv6
52	I-NLSP	Integrated Net Layer Security
53	SWIPE	IP with Encryption
54	NARP	NBMA Address Resolution Protocol
55	MOBILE	IP Mobility
56	TLSP	Transport Layer Security Protocol (using Kryptonet key management)
57	SKIP	SKIP
58	IPv6-ICMP	ICMP for IPv6
59	IPv6-NoNxt	No Next Header for IPv6
60	IPv6-Opts	Destination Options for IPv6
61		any host internal protocol
62	CFTP	CFTP
63		any local network
64	SAT-EXPAK	SATNET and Backroom EXPAK
65	KRYPTOLAN	Kryptolan
66	RVD	MIT Remote Virtual Disk Protocol
67	IPPC	Internet Pluribus Packet Core
68		any distributed file system
69	SAT-MON	SATNET Monitoring
70	VISA	VISA Protocol
71	IPCV	Internet Packet Core Utility
72	CPNX	Computer Protocol Network Executive
73	CPHB	computer Protocol Heart Beat
74	WSN	Wang Span Network
75	PVP	Packet Video Protocol
76	BR-SAT-MON	Backroom SATNET Monitoring

Exhibit 7. (Continued)   Assigned IP Numbers Field Values

Decimal	Keyword	Protocol
77	SUN-ND	SUN ND PROTOCOL-Temporary
78	WB-MON	WIDEBAND Monitoring
79	WB-EXPAK	WIDEBAND EXPAK
80	ISO-IP	ISO Internet Protocol
81	VMTP	VMTP
82	SECURE-VMTP	SECURE-VMPT
83	VINES	VINES
84	TTP	TTP
85	NSFNET-IGP	NSFNET-IGP
86	DGP	Dissimilar Gateway Protocol
87	TCF	TCF
88	EIGRP	EIGRP
89	OSPFIGP	OSPFIGP
90	Sprite-RPC	Sprite RPC Protocol
91	LARP	Locus Address Resolution Protocol
92	MTP	Multicast Transport Protocol
93	AX.25	AX.25 Frames
94	IPIP	IP-within-IP Encapsulation Protocol
95	MICP	Mobile Internetworking Control Protocol
96	SCC-SP	Semaphore Communications Sec. Protocol
97	ETHERIP	Ethernet-within-IP Encapsulation
98	ENCAP	Encapsulation Header
99		Any private encryption scheme
100	GMTP	GMTP
101	IFMP	Ipsilon Flow Management Protocol
102	PNNI	PNNI over IP
103	PIM	Protocol Independent Multicast
104	ARIS	ARIS
105	SCPS	SCPS
106	QNX	QNX
107	A/N	Active Networks
108	IPPCP	IP Payload Compression Protocol
109	SNP	Sitara Networks Protocol
110	Compaq-Peer	Compaq Peer Protocol
111	IPX-in-IP	IPX in IP
112	VRRP	Virtual Router Redundancy Protocol
113	PGM	PGM Reliable Transport protocol
114		Any 0-hop protocol
115	L2TP	Layer Two Tunneling Protocol
116	DDX	D-II Data Exchange (DDX)
117-254		Unassigned
255	Reserved	

Header Checksum Field

The header Checksum field contains a 16-bit Cyclic Redundancy Check (CRC) character. The CRC represents a number generated by treating the data in the IP header field as a long binary number and dividing that number by a fixed polynomial. The result of this operation is a quotient and remainder, with the remainder being placed into the 16-bit checksum field by the transmitting device. When a receiving station reads the header, it also performs a CRC operation on the received data, using the same field polynomial. If the computed CRC does not match the value of the CRC in the header Checksum field, the receiver assumes the header is in error and the packet is discarded. Thus, the header Checksum, as its name implies, provides a mechanism for ensuring the integrity of the IP header.

Source and Destination Address Fields

Both the Source and Destination Address fields are 32 bits in length under IPv4. The Source Address represents the originator of the datagram, while the Destination Address represents the recipient.

Under IPv4, there are five classes of IP addresses, referred to as Class A through Class E. Classes A, B, and C are subdivided into a network portion and a host portion and represent addresses used on the Internet and private IP based networks. Classes D and E represent two special types of IPv4 network addresses. Because it is extremely important to understand the composition and formation of IP addresses to correctly configure devices connected to an IP network as well as to design and modify such networks, Chapter 3 focuses on this topic.

Multiple Interface Addresses

One of the lesser-known aspects of IP addressing is the fact that it is possible to assign multiple logical network addresses to one physical network. Prior to examining how this occurs, you probably want to understand the rationale for doing this. Thus, let us assume your organization originally operated a 10BASE-5 network with 100 users and wants to construct a distributed network within a building that will consist of 350 workstations and server. Let us further assume that your organization's previously installed 10BASE-5 coaxial based backbone will be used by adding 10BASE-T hubs to the backbone, with a single router providing a connection to the Internet.

If your organization previously obtained a Class C address when it operated a 10BASE-5 network, adding 250 stations means that you would normally require a second router interface and two networks because each Class C address supports a maximum of 254 hosts. If you have no familiarity with IP addressing, like Indiana Jones, I will say "trust me" concerning host limitations until we move on to Chapter 3.

TCP/IP supports the ability to assign multiple network addresses to a common interface. In fact, TCP/IP also supports the assignment of multiple subnet numbers to a common interface. This can only be accomplished through the use of a router. Exhibit 8 illustrates an example in which three network addresses were assigned to one interface. For low volumes of network traffic, this represents an interesting technique to reduce the number of costly router interfaces required.

As indicated in Exhibit 8, the router connection to the coaxial cable would result in the assignment of two IP addresses to its interface, one for each network. In this example the addresses 205.131.175.1 and 205.131.176.1 were assigned to the router interface. Conversations between devices on the 205.131.175.0 network and the 205.131.176.0 network would require datagrams to be forwarded to the router. Thus, each station of each network would be configured with the *gateway* IP address that represents an applicable assigned router IP interface address.

Address Resolution

The TCP/IP protocol suite begins at the network layer, with an addressing scheme that identifies network address and a host address for Class A, B, and C addresses. This addressing scheme actually evolved from an ARPAnet scheme that only required hosts to be identified, since that network began as a mechanism to interconnect hosts via serial communications lines. At the same time ARPAnet was being developed, work progressed separately at the Xerox Palo Alto Research Center (PARC) on Ethernet, a technology in which multiple stations were originally connected to a coaxial cable.

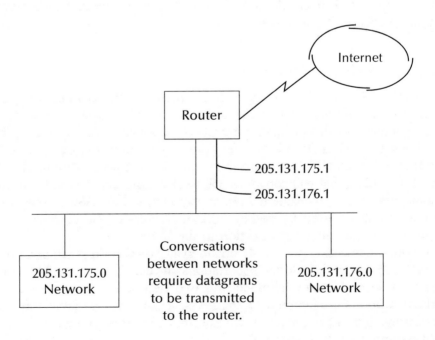

Exhibit 8.   Assigning Multiple Network Addresses to a Common Router Interface

Ethernet used a 48-bit address to identify each station on the network. As ARPAnet evolved as a mechanism to interconnect multiple hosts on geographically separated networks, IPv4 addressing evolved into a mechanism to distinguish the network and the host. Unfortunately, the addressing used by the TCP/IP protocol suite bore no relationship to the MAC address used first by Ethernet and later by Token-Ring.

Ethernet and Token-Ring Frame Formats

Exhibit 9 illustrates the frame formats for Ethernet and Token-Ring. Note that the IEEE standardized both types of LANs and uses 6-byte (48-bit) source and destination addresses. The IEEE assigns blocks of addresses 6 hex characters in length to vendors. Those 6 hex characters represent the first 24 bits of the 48-bit field used to uniquely identify a network adapter card. The vendor then encodes the remaining 24 bits or 6 hex character positions to identify the adapter card manufactured by the vendor. Thus, each Ethernet and Token-Ring adapter has a unique hardware burnt-in identifier that denotes both the manufacturer and the adapter number produced by the manufacturer.

LAN Delivery

When an IP datagram arrives at a LAN, it contains a 32-bit destination address. To deliver the datagram to its destination, the router must create a LAN frame with an appropriate MAX destination address. Thus, the router needs a mechanism to resolve or convert the IP address into the MAC address of the

Ethernet Frame

Preamble (1)	Start of Frame Delimiter (7)	Destination Address (6)	Source Address (6)	Type/ Length (2)	Information (46 to 1500)	FCS (4)

Token Ring Frame Format

Starting Delimiter (1)	Access Control (1)	Frame Control (1)	Destination Address (1)	Source Address (6)	Routing Information (Optional)

Variable Information	FCS (4)	Ending Delimiter (1)	Frame Status (1)

Legend: FCS   Frame Check Sequence
         (n)   n bytes represents field length

Exhibit 9.   Ethernet and Token-Ring Frame Formats

workstation configured with the destination IP address. In the opposite direction, a workstation may need to transmit an IP datagram to another workstation. In this situation, the workstation must be able to convert a MAC address into an IP address. Both of these address translation requirements are handled by protocols specifically developed to provide an address resolution capability. One protocol referred to as the Address Resolution Protocol (ARP) translates an IP address into a hardware address. A second protocol referred to as the Reverse Address Resolution Protocol (RARP) performs a reverse translation process, converting a hardware layer address into an IP address.

Address Resolution Operation

The address resolution operation begins when a device needs to transmit a datagram. First, the device checks its memory to determine if it previously learned the MAC address associated with a particular destination IP address. This memory location is referred to as an ARP cache. Because the first occurrence of an IP address means its associated MAC address will not be in the ARP cache, it must learn the MAC address. To do so, the device will broadcast an ARP packet to all devices on the LAN. Exhibit 10 illustrates the format of an ARP packet. Note that the numbers shown in some fields in the ARP packet indicate the byte numbers in a field when a field spans a 4-byte boundary.

ARP Packet Fields

To illustrate the operation of ARP, let us examine the fields in the ARP packet. The 16-bit Hardware Type field indicates the type of network adapter, such as 10 Mbps Ethernet (value = 1), IEEE 802 network (value = 6), and so on.

0   8   16   31

Hardware Type	Protocol Type	
Hardware Length	Protocol Length	Operation
SENDER HARDWARE ADDRESS (0–3)		
SENDER HARDWARE ADDRESS (4–5)		SENDER IP ADDRESS (0–1)
SENDER IP ADDRESS (2–3)		TARGET HARDWARE ADDRESS (0–1)
TARGET HARDWARE ADDRESS (2–5)		
TARGET IP ADDRESS		

Exhibit 10.   The ARP Packet Format

The 16-bit Protocol Type field indicates the protocol for which an address resolution process is being performed. For IP, the Protocol Type field has a value of hex 0800.

The Hardware Length field defines the number of bytes in the hardware address. Thus, the ARP packet format can be varied to accommodate different types of address resolutions beyond IP and MAC addresses. Because Ethernet and Token-Ring have the same MAC length, the value of this field is 6 for both.

The Protocol Length field indicates the length of the address for the protocol to be resolved. For IPv4 the value of this field is set to 4. The Operation field indicates the operation to be performed. This field has a value of 1 for an ARP Request. When a target station responds, the value of this field is changed to Z to denote an ARP Reply.

The Sender Hardware Address field indicates the hardware addresses of the station generating the ARP Request or ARP Reply. This field is 6 bytes in length and is followed by a 4-byte Sender IP Address field. The latter indicates the IP address of the originator of the datagram.

The next to last field is the Target Hardware Address field. Because the ARP process must discover its value, this field is originally set to all zeros in an ARP request. Once a station receives the request (and notes it has the same IP address as that in the Target IP Address field), it places its MAC address in the Target IP Address field. Thus, the last field, Target IP Address, is set to the IP address the originator needs for a hardware address.

Locating the Required Address

To put the pieces together, let us assume a router receives a datagram from the Internet with the destination IP address of 205.131.175.5. Let us further assume that the router has a connection to an Ethernet network and one station on that network has that IP address. The router needs to determine the MAC address associated with the IP address so it can construct a frame to deliver the datagram. Assuming there is no entry in its ARP cache, the router creates an ARP frame and transmits the frame using a MAC broadcast address of FFFFFFFFFFFF. Because the frame was broadcast to all stations on the network, each device reads the frame. The station that has its protocol stack configured to the same IP address as that of the Target IP Address field in the ARP frame would respond to the ARP Request. When it does, it will transmit an ARP Reply in which its physical MAC address is inserted into the ARP Target Hardware Address field that was previously set to zero.

The ARP standard includes provisions for devices on a network to update their ARP table with the MAC and IP address pair of the sender of the ARP Request. Thus, as ARP Requests flow on a LAN, they contribute to the building of tables that reduce the necessity of additional broadcasts.

Gratuitous ARP

There is a special type of ARP referred to as a *gratuitous ARP* that deserves mention. When a TCP/IP stack is initialized, it issues a gratuitous ARP, which represents an ARP request for its own IP address. If the station receives a reply containing a MAC address that differs from its address, this indicates that another device on the network is using its assigned IP address. If this situation occurs, an error message warning you of an address conflict will be displayed.

Proxy ARP

A proxy is a device that works on the behalf of another. Thus, a proxy ARP represents a mechanism, which enables a device to answer an ARP request on behalf of another device.

The rationale for the development of proxy ARP, which is also referred to as ARP Hack, dates to the early use of subnetting when a LAN could be subdivided into two or more segments. If a station on one segment required the MAC address of a station on another subnet, the router would block the ARP request because it is a Layer 2 broadcast, and routers operate at Layer 3. Because the router is aware of both subnets, it could answer an ARP request on one subnet on behalf of other devices on the second subnet by supplying its own MAC address. The originating device will then enter the router's MAC address in its ARP cache and will correctly transmit packets destined for the end host to the router.

RARP

The Reverse Address Resolution Protocol (RARP) was at one time quite popular when diskless workstations were commonly used in corporations. In such situations, the workstation would know its MAC address, but be forced to learn its IP address from a server on the network. Thus, the RARP protocol would be used by the client to access a server on the local network and would provide the client's IP address. Similar to ARP, RARP is a Layer 2 protocol that cannot normally cross router boundaries. Some router manufacturers implemented RARP, which allows requests and responses to flow between networks.

The RARP frame format is the same as ARP. The key difference between the two is the setting of field values. The RARP protocol fills in the sender's hardware address and sets the IP address field to zeros. Upon receipt of the RARP frame, the RARP server fills in the IP address field and transmits the frame back to the client, reversing the ARP process.

ICMP

To conclude this section, attention will turn to the Internet Control Message protocol (ICMP). If we think about the IP protocol for a while, we would note that there is no provision to inform a source of the fact that a datagram encountered some type of problem. This is because one of the functions of

ICMP is to provide a messaging capability that reports different types of errors that can occur during the processing of datagrams. In addition to providing an error reporting mechanism, ICMP includes certain types of messages that provide a testing capability.

Overview

ICMP messages are transmitted within an IP datagram as illustrated in Exhibit 11. Note that although each ICMP message has its own format, they all begin with the same three fields. Those fields are an 8-bit Type field, an 8-bit Code field, and a 16-bit Checksum field.

We can obtain familiarity with the capability of ICMP by examining the use of some of the fields within an ICMP message. Thus, let us turn our attention to the Type and Code fields within an ICMP message.

The ICMP Type Field

The purpose of the ICMP Type field is to define the meaning of the message as well as its format. Two of the most popularly used ICMP messages use type values of 0 and 8. A Type field value of 8 represents an echo request, while a Type field value of 0 denotes an ECMP echo reply. Although their official names are Echo Request and Echo Reply, most persons are more familiar with the term Ping that is used to reference both the request and the reply.

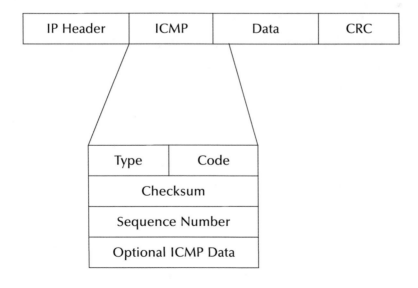

Exhibit 11. Transporting ICMP Messages via Encapsulation within an IP Datagram

Exhibit 12 lists ICMP Type field values that currently identify specific types of ICMP messages.

Exhibit 12.   ICMP Type Field and Code Values

Type 0 Echo Reply Message
ICMP Fields:

Type

    0 for echo reply message.

Code

    0

    Code 0 may be received from a gateway or a host.

Type 3 Destination Unreachable Message
ICMP Fields:

Type

    3

Code

    0 = net unreachable;

    1 = host unreachable;

    2 = protocol unreachable;

    3 = port unreachable;

    4 = fragmentation needed and DF set;

    5 = source route failed.

    Codes 0, 1, 4, and 5 may be received from a gateway. Codes 2
    and 3 may be received from a host.

Type 4 Source Quench Message
ICMP Fields:

Type

    4

Code

    0

    Code 0 may be received from a gateway or a host.

Type 5 Source Redirect Message
ICMP Fields:

Type

    5

Code

    0 = Redirect datagrams for the Network.

    1 = Redirect datagrams for the Host.

    2 = Redirect datagrams for the Type of Service and Network.

    3 = Redirect datagrams for the Type of Service and Host.

    Codes 0, 1, 2, and 3 may be received from a gateway.

Exhibit 12. (Continued)   ICMP Type Field and Code Values

Type 8 Echo
ICMP Fields:
   Type
      8 for echo message;
   Code
      0

Type 11 Time Exceeded Message
ICMP Fields:
   Type
      11
   Code
      0 = time to live exceeded in transit;
      1 = fragment reassembly time exceeded.

      Code 0 may be received from a gateway. Code 1 may be received
      from a host.

Type 12 Parameter Problem Message
ICMP Fields:
   Type
      12

      Code 0 may be received from a gateway or a host.

Type 13 Timestamp
ICMP Fields:
   Type
      13 for timestamp message;
   Code
      0

Type 14 Timestamp Reply Message
ICMP Fields:
   Type
      14 for timestamp reply message.
   Code
      0

Type 15 Information Request
ICMP Fields:
   Type
      15 for information request message
   Code
      0

Type 16 Information Reply Message
ICMP Fields:
   Type
      16 for information reply message.
   Code
      0

The ICMP Code Field

The ICMP Code field provides additional information about a message defined in the Type field. The Code field may not be meaningful for certain ICMP messages. For example, both Type field values of 0 (echo reply) and 8 (echo request) always have a Code field value of 0. In comparison, a Type field value of 3 (Destination Unreachable) can have 1 of 16 possible Code field values, which further defines the problem. Exhibit 13 lists the Code field values presently assigned to ICMP messages based on their Type field values.

Evolution

Over the years from its first appearance in RFC 792, ICMP has evolved through the addition of many functions. For example, a Type 4 (source quench) represents the manner by which an end station indicates to the originator of a message that the host cannot accept the rate at which the originator is transmitting packets. The recipient will send a flow of ICMP Type 4 messages to the originator as a message for the origination to slow down its transmission. When an acceptable flow level is reached, the recipient will terminate its generation of source quench messages. Although popularly used many years ago for controlling traffic, the TCP slow-start algorithm has superseded a majority of the use of ICMP Type 4 messages.

ICMP message types that warrant discussion are Type 5 and Type 7. A router generates a Type 5 (redirect) message when it receives a datagram and determines that there is a better route to the destination network. This ICMP message informs the sender of the better route. A Type 7 message (time exceeded) indicated that the time to live field value in an IP datagram header was decremented to 0, and the datagram was discarded. As we will note later in this book, ICMP provides a foundation for several diagnostic testing application. Unfortunately, this testing capability can be abused by unscrupulous persons and results in many organizations filtering ICMP messages so they do not flow from the Internet onto a private network.

The Transport Layer

The purpose of this section is to acquaint you with the two transport layer protocols supported by the TCP/IP suite. Those protocols are the Transmission Control Protocol (TCP) and the User Datagram Protocol (UDP).

As indicated in the section entitled The Internet Protocol and Related Protocols, either TCP or UDP can be identified by the setting of an applicable value in the IP Header. Although the use of either protocol results in the placement of the appropriate transport layer header behind the IP header, there are significant differences between the functionality of each transport protocol. Those differences make one protocol more suitable for certain applications than the other protocol and vice versa.

Exhibit 13. ICMP Code Field Values Based on Message Type

Message Type	Code Field Values
3	Destination Unreachable Codes
	0 Net Unreachable
	1 Host Unreachable
	2 Protocol Unreachable
	3 Port Unreachable
	4 Fragmentation Needed and Don't Fragment was Set
	5 Source Route Failed
	6 Destination Network Unknown
	7 Destination Host Unknown
	8 Source Host Isolated
	9 Communication with Destination Network is Administratively Prohibited
	10 Communication with Destination Host is Administratively Prohibited
	11 Destination Network Unreachable for Type of Service
	12 Destination Host Unreachable for Type of Service
	13 Destination Host Unreachable for Type of Service
	14 Communication Administratively Prohibited
	15 Precedence cutoff in effect
5	Redirect Codes
	0 Redirect Datagram for the Network (or subnet)
	1 Redirect Datagram for the Host
	2 Redirect Datagram for the Type of Service and Network
	3 Redirect Datagram for the Type of Service and Host
6	Alternate Host Address Codes
	0 Alternate Address for Host
11	Time Exceeded Codes
	0 Time to Live Exceeded in Transit
	1 Fragment Reassembly Time Exceeded
12	Parameter Problem Codes
	0 Point Indicates the Error
	1 Missing a Required Option
	2 Bad Length
40	Photuris Codes
	0 Reserved
	1 Unknown Security Parameters Index
	2 Valid Security Parameters, but Authentication Failed
	3 Valid Security Parameters, but Decryption Failed

TCP

The Transmission Control Protocol is a connection-oriented protocol. This means that the protocol will not forward data until a session is established in which the destination acknowledges it is ready to receive data. This also means that the TCP setup process requires more time than when UDP is used as the transport layer protocol. However, because you would not wish to commence certain operations such as remote log-on or a file transfer unless you knew the destination was ready to support the appropriate application, the use of TCP is more suitable for certain applications than UDP. Conversely, when we examine UDP, we will note that this transport layer protocol similarly supports certain applications between than other applications. Because the best way to become familiar with TCP is by first examining the fields in its header, we will do so.

The TCP Header

The TCP header consists of 12 fields as illustrated in Exhibit 14. When examining the UDP header later in the chapter, we will note by comparing the two that the TCP header is far more complex. The reason for this additional complexity results from the fact that TCP is not only a connection-oriented protocol, but, in addition, supports error detection and correction as well as packet sequencing, with the latter used to note the ordering of packets to include determining if one or more packets are lost.

Source Port								Destination Port	
Sequence Number									
Acknowledgment Number									
Hlen	Reserved	URG	ACK	PSH	RST	SYN	FIN	Window	
Checksum								Urgent Pointer	
Options								Padding	

Exhibit 14.   The TCP Header

Source and Destination Port Fields

The Source and Destination port fields are each 16 bits in length. Each field denotes a particular process or application. In actuality, most applications use the destination port number to denote a particular process or application and either set the Source port field value to a random number greater than 1024 or to zero. The reason the destination port number defines the process or application results from the fact that an application operating at the receiver normally operates acquiescently, waiting for requests, looking for a specific destination port number to determine the request.

The reason the originator sets the source port to 0 or a value above 1023 is due to the fact that the first 1023 out of 65,536 available port numbers are standardized with respect to the type of traffic transported via the use of specific numeric values. To illustrate the use of port numbers, assume one station wishes to open a Telnet connection with a distant server. Because Telnet is defined as port 23, the application will set the destination port value to that numeric. The source port will normally be set to a random value above 1023 and an IP header will then add the destination and source IP addresses for routing the datagram from the client to the server. In some literature you may encounter the term *socket*, sometimes incorrectly used as a synonym for port. In actuality the destination port in the TCP or UDP header plus the destination IP address cumulatively identify a unique process or application on a host. The combination of port number and IP address is correctly referenced as a socket. At the server, the destination port value of 23 identifies the application as Telnet. When the server forms a response, it first reverses source and destination IP addresses. Similarly, the server places the source port number in the destination port field, which enables the Telnet originator's application to correctly identify the response to its initial datagram.

Multiplexing and Demultiplexing

Port numbers play an important role in TCP/IP because they enable multiple applications to flow between the same pair of stations or from multiple nonrelated stations to a common station. This flow of multiple applications to a common address is referred to as multiplexing. Upon receipt of a datagram, the removal of the IP and TCP headers requires the remaining portion of the packet to be routed to its correct process or application based on the destination port number in the TCP header. This process is referred to as demultiplexing.

Both TCP and UDP use port numbers to support the multiplexing of different processes or applications to a common IP address. An example of this multiplexing and demultiplexing of packets is illustrated in Exhibit 15.

The top left portion of Exhibit 15 illustrates how both Telnet and FTP, representing two TCP applications, could be multiplexed into a stream of IP datagrams that flow to a common IP address. In comparison, the top right portion of Exhibit 15 illustrates how, through port numbering, UDP ports permit a similar method of multiplexing of applications.

Port Numbers

The *universe* of both TCP and UDP port numbers can vary from a value of 0 to 65535, resulting in a total of 65536 ports capable of being used by each

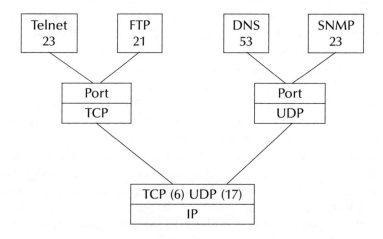

Exhibit 15.   Multiplexing Multiple Applications via Serial Communications to a Common IP Address

transport protocol. This so-called port universe is divided into three ranges referred to as Well-Known ports, Registered ports, and Dynamic or Private ports.

- *Well-Known ports* — The most commonly used port values because they represent assigned numeric that identify specific processes or applications. Ports 0 through 1023 represent the range of Well-Known ports. These port numbers are assigned by the Internet Assigned Numbers Authority (IANA) and are used to indicate the transportation of standardized processes. When possible, the same Well-Known port number assignments are used with TCP and UDP. Ports used with TCP are normally used to provide connections that transport long-term conversations. In some literature, you may encounter Well-Known port numbers being specified as in the range of value from 0 to 255. While this range was correct many years ago, the modern range for assigned ports managed by the IANA were expanded to cover the first 1024 port values from 0 to 1023. Exhibit 16 provides a summary of the port value assignments from 0 through 255 for Well-Known ports to include the service supported by a particular port and the type of port, TCP or UDP, for which the port number is primarily used. A good source for the full list of assigned port numbers is RFC 1700.
- *Registered ports* — Ones whose values range from 1024 through 49151. Although all ports above 1023 can be used freely, the IANA requests vendors to register their application port numbers with them.
- *Dynamic or private ports* — The third range of port numbers is from 49152 through 65535. This port number range is associated with dynamic or private ports. This port range is usually used by new applications that remain to be standardized, such as Internet telephony.

Exhibit 16. Well-Known TCP and UDP Services and Port Use

Keyword	Service	Port Type	Port Number
TCPMUX	TCP Port Service Multiplexer	TCP	1
RJE	Remote Job Entry	TCP	5
ECHO	Echo	TCP and UDP	7
DAYTIME	Daytime	TCP and UDP	13
QOTD	Quote of the Day	TCP	17
CHARGEN	Character Generator	TCP	19
FTD-DATA	File Transfer (Default Data)	TCP	20
FTP	File Transfer (Control)	TCP	21
TELNET	Telnet	TCP	23
SMTP	Simple Mail Transfer Protocol	TCP	25
MSG-AUTH	Message Authentication	TCP	31
TIME	Time	TCP	37
NAMESERVER	Host Name Server	TCP and UDP	42
NICNAME	Who Is	TCP	43
DOMAIN	Domain Name Server	TCP and UDP	53
BOOTPS	Bootstrap Protocol Server	TCP	67
BOOTPC	Bootstrap Protocol Client	TCP	68
TFTP	Trivial File Transfer Protocol	UDP	69
FINGER	Finger	TCP	79
HTTP	World Wide Web	TCP	80
KERBEROS	Kerberos	TCP	88
RTELNET	Remote Telenet Service	TCP	107
POP2	Post Office Protocol Version 2	TCP	109
POP3	Post Office Protocol Version 3	TCP	110
NNTP	Network News Transfer Protocol	TCP	119
NTP	Network Time Protocol	TCP and UDP	123
NETBIOS-NS	NetBIOS Name Server	UDP	137
NETBIOS-DGM	NetBIOS Datagram Service	UDP	138
NETBIOS-SSN	NetBIOS Session Service	UDP	139
NEWS	News	TCP	144
SNMP	Simple Network Management Protocol	UDP	161
SNMTTRAP	Simple Network Management Protocol Traps	UDP	162
BGP	Border Gateway Protocol	TCP	179
HTTPS	Secure HTTP	TDP	413
RLOGIN	Remote Login	TCP	513
TALK	Talk	TCP and UDP	517

Sequence and Acknowledgment Number Fields

TCP is a byte-oriented sequencing protocol. Thus, a Sequence field is necessary to ensure that missing or misordered packets are noted or identified. That field is 32 bits in length and provides the mechanism for ensuring that missing or misordered packets are noted or identified.

The actual entry in the Sequence Number field is based on the number of bytes in the TCP data field, i.e., because TCP was developed as a byte-oriented protocol, each byte in each packet is assigned a sequence number. Because it would be most inefficient for TCP to transmit 1 byte at a time, groups of bytes, typically 512 or 536, are placed in a segment and 1 sequence number is assigned to the segment and placed in the sequence field. That number is based on the number of bytes in the current segment as well as previous segments, because the sequence field value increments its count until all 16-bit positions are used and then continues via a rollover through zero. For example, assume the first TCP segment contains 512 bytes and a second segment will have the sequence number 1024.

The Acknowledgment Number field, which is also 32 bits in length, is used to verify the receipt of data. The number in this field also reflects bytes. For example, returning to our sequence of two 512-byte segments, when the first segment is received, the receiver expects the next sequence number to be 513. Therefore, if the receiver were acknowledging each segment, it would first return an acknowledgment with a value of 513 in the Acknowledgment Number field. When it acknowledges the next segment, the receiver would set the value in the acknowledgment number field to 1025 and so on.

Because it would be inefficient to have to acknowledge each datagram, a variable or "sliding" window is supported by the TCP protocol, i.e., returning an acknowledgment number field value of n + 1 would indicate the receipt of all bytes through byte n. If the receiver has the ability to process a series of multiple segments and each is received without error, it would be less efficient to acknowledge each datagram. Thus, a TCP receiver can process a variable number of segments prior to returning an acknowledgment that informs the transmitter that n bytes were received correctly. To ensure lost datagrams or lost acknowledgments do not place the TCP protocol in an infinite waiting period, the originator sets a timer and will retransmit data if it does not receive a response within a predefined period of time.

The previously described use of the Acknowledgment Number field is referred to as Positive Acknowledgment Retransmission (PAR). Under PAR, each unit of data must be either implicit (sending a value of n + 1 to acknowledge receipt of n bytes) or explicit. If a unit of data is not acknowledged by the time the originator's time out period is reached, the previous transmission is retransmitted. When the Acknowledgment Number field is in use, a flag bit, referred to as the ACK flag in the Code field, will be set. We will shortly discuss the six bit positions in the Code bit field.

Hlen Field

The Header Length (Hlen) field is 4 bits in length. This field, which is also referred to as the Offset field, contains a value that indicates where the TCP header ends and the data field begins. This value is specified as a number of 32-bit words. It is required due to the fact that the inclusion of options can result in a variable length header. Because the minimum length of the TCP header is 20 bytes, the minimum value of the Hlen field would be 5, denoting five 32-bit words which equals 20 bytes.

Code Bits Field

As previously indicated in Exhibit 14, there are six individual 1-bit fields within the Code bits field. Each bit position functions as a flag to indicate whether or not a function is enabled or disabled. Thus, to obtain an appreciation for the use of the Code bits field, we need to examine each bit position in that field — which we will now do.

- *URG bit* — The Urgent (URG) bit or flag is used to denote an urgent or priority activity. When such a situation occurs, an application will set the URG bit position, which acts as a flag and results in TCP immediately transmitting everything it has for the connection instead of waiting for additional characters. An example of an action that could result in an application setting the urgent flag would be a user pressing the CTRL-BREAK key combination. A second meaning resulting from the setting of the urgent bit or flag is that it also indicates the Urgent Pointer field is in use. Here the Urgent Pointer field indicates the offset in bytes from the current sequence number where the urgent data is located.
- *ACK bit* — The setting of the ACK bit indicates that the segment contains an acknowledgment to a previously transmitted datagram or series of datagrams. Then the value in the acknowledgment number field indicates the correct receipt of all bytes through byte n by having the byte number n + 1 in the field.
- *PSH bit* — The third bit position in the Code bit field is the Push (PSH) bit. This one bit field is set to request the receiver to immediately deliver data to the application and flags any buffering. Normally the delivery of urgent information would result in the setting of both the URG and PSH bits in the Code Bit field.
- *RST bit* — The fourth bit position in the Code bit field is the reset (RST) bit. This bit position is set to reset a connection. By responding to a connection request with the RST bit set, this bit position can also be used as a mechanism to decline a connection request.
- *SYN bit* — The fifth bit in the Code bit field is the Synchronization (SYN) bit. This bit position is set at start-up during what is referred to as a three-way handshake, which will be covered shortly.
- *FIN bit* — The sixth and last bit position in the Code bit field is the finish (FIN) bit. This bit position is set by the sender to indicate that it has no additional data, and the connection should be released.

Window Field

The Window field is 16 bits in length and provides TCP with the ability to regulate the flow of data between source and destination. Therefore, this field indirectly performs flow control.

The Window field indicates the maximum number of bytes that the receiving device can accept. Thus, it indirectly indicates the available buffer memory of the receiver. Here a large value can significantly improve TCP performance because it permits the originator to transmit a number of segments without having to wait for an acknowledgment while permitting the receiver to acknowledge the receipt of multiple segments with one acknowledgment.

Because TCP is a full-duplex transmission protocol, both the originator and recipient can insert values in the Window field to control the flow of data in each direction. By reducing the value in the Window field, one end of a session in effect informs the other end to transmit less data. Thus, the use of the Window field provides a bi-directional flow control capability.

Checksum Field

The Checksum field is 16 bits or 2 bytes in length. The function of this field is to provide an error detection capability for TCP. To do so, this field is primarily concerned with ensuring that key fields are validated instead of protecting the entire header. Thus, the Checksum calculation occurs over what is referred to as a 12-byte pseudo header. This pseudo header includes the 32-bit Source and Destination Address fields in the IP header, the 8-bit protocol field, and a length field that indicates the length of the TCP header and data transported within the TCP segment. Thus, the primary purpose of the Checksum field is to ensure data arrived at its correct destination, and the receiver has no doubt about the address of the originator nor the length of the header and the type of application data transported.

Urgent Pointer Field

The Urgent Pointer field is 1 byte in length. The value in this field acts as a pointer to the sequence number of the byte following the urgent data. As previously noted, the URG bit position in the Code field must be set for the data in the Urgent Pointer field to be interpreted.

Options

The Options field, if present, can be variable in length. The purpose of this field is to enable TCP to support various options, with Maximum Segment Size (MSS) representing a popular TCP option. Because the header must end on a 32-bit boundary, any option that does not do so is extended via pad characters that in some literature is referred to as a Padding field.

Padding Field

The Padding field is optional and is included only when the Options field does not end on a 32-bit boundary. Thus, the purpose of the Padding field is to ensure that the TCP header, when extended, falls on a 32-bit boundary. Now that we have an appreciation for the composition of the TCP header, let us turn our attention to the manner by which this protocol operates. In doing so, we will examine how TCP establishes a connection with a distant device and its initial handshaking process, its use of sequence and acknowledgment numbers, how flow control is supported by the protocol, and how the protocol terminates a session.

Connection Establishment

As mentioned earlier in this section, TCP is a connection-oriented protocol that requires a connection between two stations to be established prior to the actual transfer of data occurring. The actual manner by which an application communicates with TCP is through a series of function calls. In order to obtain an appreciation for the manner by which TCP establishes a session, we must first examine connection function calls used by applications, for example, Telnet and FTP.

Connection Function Calls

Exhibit 17 illustrates the use of the OPEN connection function calls during the TCP connection establishment process. This process commences when an

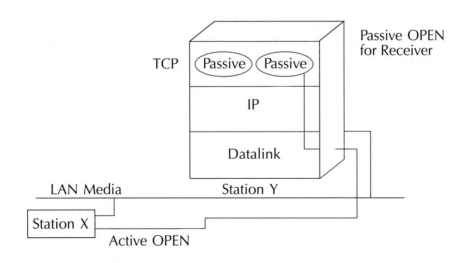

Exhibit 17.   Using Function Calls to Establish a TCP Connection

application requires a connection to a remote station. At that time the application will request TCP to place an OPEN function call. There are two types of OPEN function calls, referred to as (1) passive and (2) active. A passive OPEN function call represents a call to allow connections to be accepted from a remote station. This type of call is normally issued upon application startup, informing TCP that, for example, FTP or Telnet is active and ready to accept connections originating from other stations. TCP will then note that the application is active and will also note its port assignment. The TCP protocol will then allow connections on that port number.

- *Port hiding* — One of the little known aspects of TCP is the fact that some organizations attempt to hide their applications by configuring applications for ports other than well-known ports. For example, assigning Telnet to port 2023 instead of port 23 is an example of port hiding. Although a person with port scanning software would be able to easily discover that port 2023 is being used, the theory behind port hiding is that it reduces the ability of lay personnel to easily discover applications at different network addresses and then attempt to use those applications.

- *Passive OPEN* — Returning to the use of a passive OPEN function call, its use governs the number of connections allowed, i.e., while a client would usually issue one passive OPEN, a server would issue multiple OPENs since it is designed to service multiple sessions. Another term used for the passive end of the TCP action is responder or TCP responder. Thus, a TCP responder can be thought of as an opening up of connection slots to accept any inbound connection request without waiting for any particular station request.

- *Active OPEN* — A station that needs to initiate a connection to a remote station issues the second type of OPEN call. This type of function call is referred to as an active OPEN. In the example illustrated in Exhibit 17, station X would issue an active OPEN call to station Y. For the connection to be serviced by station Y, that station must have previously issued a passive OPEN request which, as previously explained, allows incoming connections to be established. To successfully connect, station X's active OPEN must use the same port number that the passive OPEN used on station Y. In addition to active and passive OPEN calls, other calls include CLOSE (to close a connection), SEND and RECEIVE (to transfer information), and STATUS (to receive information for a previously established connection).

Now that we have an appreciation for the use of active and passive OPEN calls to establish a TCP connection, turn to the manner by which TCP segments are exchanged. As we will note, the exchange of segments enables a session to occur. The initial exchange of datagrams that transport TCP segments is referred to as a three-way handshake. It is important to note how and why this process occurs. It has been used in modified form as a mechanism to create a denial of service (DOS) attack, which we will examine in Chapter 9.

The Three-Way Handshake

To ensure that the sender and receiver are ready to commence the exchange of data requires both parties to the change to be synchronized. Thus, during the TCP initialization process, sender and receiver exchange a few control packets for synchronization purposes. This exchange is referred to as a three-way handshake. This functions as a mechanism to synchronize each endpoint at the beginning of a TCP connection with a sequence number and an acknowledgment number.

As we will soon note, a three-way handshake begins with the originator sending a segment with its SYN bit in the Code bit field set. The receiving station will respond with a similar segment with its ACK bit in the Code bit field set. Thus, an alternate name for the three-way handshake is an *initial SYN-SYN-ACK* sequence.

Operation

To illustrate the three-way handshake, let us continue from our prior example shown in Exhibit 17, in which station X placed an active OPEN call to TCP to request a connection to a remote station and an application on that station. Once the TCP/IP protocol stack receives an active OPEN call, it will construct a TCP header with the SYN bit in the Code bit field set. The stack will also assign an initial sequence number and place that number in the Sequence Number field in the TCP header. Other fields in the header, such as the destination port number, are also set and the segment is then transferred to IP for the formation of a datagram for transmission onto the network.

To illustrate the operation of the three-way handshake, consider Exhibit 18 which illustrates the process between stations X and Y. Because the initial sequence number does not have to start at zero, we will assume it commenced at 1000 and then further assume that the value was placed in the Sequence number field. Thus, the TCP header flowing from station X to station Y is shown with SYN=1 and SEQ=1000.

Because the IP header results in the routing of a datagram to station Y, that station strips the IP header and notes that the setting of the SYN bit in the TCP header represents a connection request. Assuming station Y can accept a new connection, it will acknowledge the connection request by building a TCP segment. That segment will have its SYN and ACK bits in its Code bit field set. In addition, station Y will place its own initial sequence number in the Sequence Number field of the TCP header it is forming. Because the connection request had a sequence number of 1000, station Y will acknowledge receipt by setting its Acknowledgment field value to 1001 (station X sequence number plus 1) which indicates the next expected sequence number.

Once station Y forms its TCP segment, the segment has an IP header added to form a datagram. The datagram flows to station X. Station X receives the

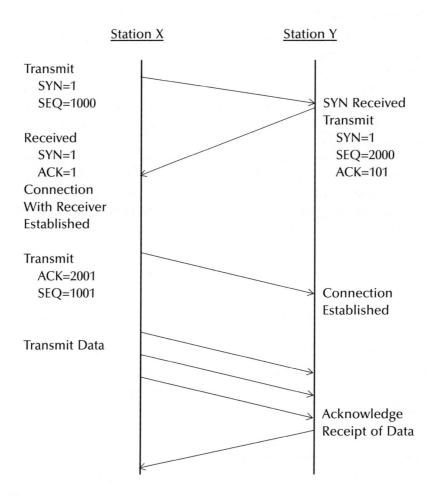

Exhibit 18.   The Three-Way Handshake

datagram, removes the IP header and notes via the setting of the XYN and ACK bits and Sequence Number field value that it is a response to its previously issued connection request. To complete the connection request, station X must, in effect, acknowledge the acknowledgment. To do so, station X will construct a new TCP segment in which the ACK bit will be set and the sequence number will be incremented by 1 to 1001. Station X will also set the acknowledgment number to 2001 and form a datagram that is transmitted to station Y. Once station Y examines the TCP header and confirms the correct values for the Acknowledgment and Sequence Number fields, the connection becomes active. At this point in time, both data and commands can flow between the two endpoints. As this occurs, each side of the connection maintains its own set of tables for transmitted and received sequence numbers. Those numbers are always in ascending order. When the applicable 16-bit field reaches its maximum value, the settings wrap to 0.

In examining the three-way handshake illustrated in Exhibit 18, note that after the originating station establishes a connection with the receiver, it transmits a second TCP initialization segment to the receiver, and follows that segment with one or more IP datagrams that transport the actual data. In Exhibit 18, a sequence of three datagrams is shown being transmitted prior to station Y generating an acknowledgment to the three segments transported in the three datagrams. The actual number of outstanding segments depends on the TCP window, so our attention will turn to this topic.

The TCP Window

TCP is a connection-oriented protocol that includes a built-in capability to regulate the flow of information, a function referred to as flow control. TCP manages the flow of information by increasing or decreasing the number of segments that can be outstanding at any point in time. For example, under periods of congestion when a station is running out of available buffer space, the receiver may indicate it can only accept one segment at a time and delay its acknowledgment to ensure it can service the next segment without losing data. Conversely, if a receiver has free and available buffer space, it may allow multiple segments to be transmitted to it and quickly acknowledge the segments.

TCP forms segments sequentially in memory. Each segment of memory waits for an IP header to be added to form a datagram for transmission. A "window" is placed over this series of datagrams that structures three types of data: (1) data transmitted and acknowledged, (2) data transmitted but not yet acknowledged, and (3) data waiting to be transmitted. Because this "window" slides over the three types of data, the window is referred to as a *sliding window.*

Exhibit 19 illustrates the use of the TCP sliding window for flow control purposes. Although the actual TCP segment's size is normally 512 bytes, for simplicity of illustration, a condensed sequence of segments with sequence numbers varying by unity are shown. In this example, we will assume that

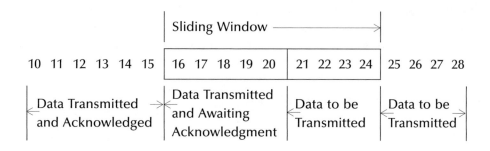

Exhibit 19.   Flow Control and the TCP Sliding Window

sequence numbers 10 through 15 have been transmitted to the destination station. The remote station acknowledges receipt of those segments. Datagrams containing segment sequence numbers 16 through 20 were transmitted by the source station, but at this particular point in time have not received an acknowledgment. Thus, that data represents the second type of data covered by a sliding window. Note that this window will slide up the segments as each datagram is transmitted. The third type of data covered by the siding window is segments. In Exhibit 19 segments 21 through 24 are in the source station awaiting transmission, while segments 25 through 28 are awaiting coverage by the sliding window.

If we return to Exhibit 14, which illustrated the TCP header, we will note a field labeled *Window*. That field value indirectly governs the length of the sliding window. In addition, the setting of that field provides a flow control mechanism. For example, the Window field transmitted by a receiver to a sender indicates the range of sequence number, which equates to bytes, that the receiver is willing to accept. If a remote station cannot accept an additional data, it would then set the Window field value to zero. The receiving station continues to transmit TCP segments with the Window field set to zero until its buffer is emptied a bit (no pun intended), in effect allowing the resumption of transmission conveying data by the originator. That is, when the transmitting station receives a response with a Window field value of zero, it replies to the response with an ACK (Code field ACK bit set to one) and its Window field set to a value of zero. This inhibits the flow of data. When sufficient buffer space becomes available at the receiver, it will form a segment with its Window field set to a non-zero value, indicating that it can again receive data. At this point, the transmitting of data goes to the receiver.

Avoiding Congestion

One of the initial problems associated with TCP is the fact that a connection could commence with the originator transmitting multiple segments, up to the Window field value returned by the receiver during the previously described three-way handshake. If there are slow-speed WAN connections between originator and recipient, it is possible that routers could become saturated when a series of transmissions originated at the same time. In such a situation, the router would discard datagrams, causing retransmissions that continued the abnormal situation. The solution developed to avoid this situation is referred to as a TCP slow start process.

TCP Slow Start

Slow start represents an algorithm procedure added to TCP that implicitly uses a new window, referred to as the congestion window. This window is not contained as a field in the TCP header. Instead it becomes active through the algorithm that defined the slow start process. That is, when a new connection is established, the congestion window is initialized to a size of one segment, typically 512 or 536 bytes. Each time an ACK is received, the congestion window's length is increased by one segment. The originator can transmit any number of segments up to the minimum value of the congestion window or the Window field value (Advertised Window). Note that flow control is imposed by the transmitter in one direction through the congestion window, while it is imposed in the other direction by the receiver's advertised Window field value.

Although slow start commences with a congestion window of one segment, it builds up exponentially until it reaches the Advertised Window size, i.e., it is incremented by subsequent ACKs from 1 to 2, then it is increased to 4, 8, 16, and so on, until it reaches the Advertised Window size. Once this occurs, segments are transferred using the Advertised Window size for congestion control and the slow start process is terminated.

The Slow Start Threshold

In addition to working at initiation, slow start will return on the occurrence of one of two conditions: (1) duplicate ACKs or (2) a time out condition where a response is not received within a predefined period of time. When either situation occurs, the originator commences another algorithm referred to as the Congestion Control Algorithm.

When congestion occurs, a comparison is initiated between the Congestion Window size and the current Advertised Window size. The smaller number is halved and saved in a variable referred to as a slow start threshold. The minimum value of the slow start threshold is two segments unless congestion occurred via a time out, with the congestion window then set to a value of 1, the same as a slow start process. The TCP originator has the option of using the slow start start-up or congestion avoidance. To determine which method to use, the originator compares the congestion value to the value of the slow start threshold. If the congestion value matches the value of the slow start threshold, the congestion avoidance algorithm will be used. Otherwise, the originator will use the slow start method. Since we previously described the slow start method, let us turn to the congestion avoidance method and to the algorithm it uses.

Upon the receipt of ACKs, the congestion window will be increased until its value matches the value saved in the slow start threshold. When this occurs, the slow start algorithm terminates and the congestion avoidance algorithm starts. This algorithm multiplies the segment size by two, divides that value by the congestion window size, and then continually increases its value based on the previously described algorithm each time an ACK is received. The result of this algorithm is a more linear growth in the number of segments that can be transmitted in comparison to the exponential growth of the slow start algorithm.

TCP Retransmissions

While it is obvious that the negative acknowledgment of a segment by the receiver returning the same segment number expected indicates a retransmission request, what happens if a datagram is delayed? Since delays across a TCP/IP network depend on the activity of other routers in the network, the number of hops in the path between source and destination, and other factors, it is relatively impossible to have an exact expected delay prior to a station assuming data is lost and retransmitting. Recognizing this situation, the developers of TCP included an adaptive retransmission algorithm in the protocol. Under this algorithm, when TCP submits a segment for transmission, the protocol records the segment sequence number and time. When an acknowledgment is received to that segment, TCP also records the time, obtaining a round-trip delay. The TCP protocol uses such timing information to construct an average round-trip delay that is used by a timer to denote, when the timer expires, that a retransmission should occur. When a new transmit-response sequence occurs, another round-trip delay is computed which slightly changes the average. Thus, this technique slowly changes the timer value that governs the acceptable delay for waiting for an ACK. Now that we have an appreciation for the manner by which TCP determines when to retransmit a segment, let us conclude our coverage of this protocol by turning to the manner by which it gracefully terminates a session.

Session Termination

If we remember the components of the Code bit field, we previously noted that field has a FIN bit. The purpose of this bit is to enable TCP to gracefully terminate a session. Before TCP supports full-duplex communication, each party to the session must close the session. This means that both the originator and recipient must change segments with the FIN bit set in each segment.

Exhibit 20 illustrates the exchange of segments to gracefully terminate a TCP connection. In this example, assume station X has completed its transmission and indicates this fact by sending a segment to station Y with the FIN bit set. Station Y will acknowledge the segment with an ACK. At this point in time, station Y will no longer accept data from station X. Station Y can continue to accept data from its application to transmit to station X. If station Y does not have any more data to transmit, it will then completely close the connection by transmitting a segment to station X with the FIN bit set in the segment. Station X will then ACK that segment and terminate the connection. If an ACK should be lost in transit, segments with FIN are transmitted and a timer is set. Then either an ACK is received or a time-out occurs which serves to close the connection.

UDP

The User Datagram Protocol (UDP) is the second transport layer protocol supported by the TCP/IP protocol suite. UDP is a connectionless protocol.

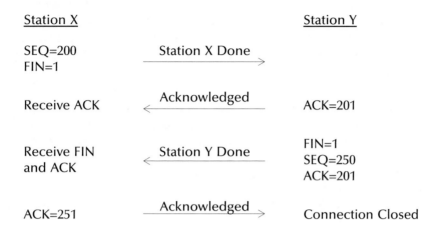

Exhibit 20. Terminating a TCP Connection

This means that an application using UDP can have its data transported in the form of IP datagrams without first having to establish a connection to the destination. This also means that when transmission occurs via UDP, there is no need to release a connection, simplifying the communication's process. Other features of UDP include the fact that this protocol has no ordering capability nor does it provide an error detection and correction capability. This in turn results in a header that is greatly simplified and is much smaller than TCPs.

The UDP Header

Exhibit 21 illustrates the composition of the UDP header. This header consists of 64 bytes followed by actual user data. In comparing the TCP and UDP headers, it is easy to note the relative simplicity of the latter since it lacks many of the features of the former. For example, because it does not require the acknowledgment of datagrams nor sequence datagrams, there is no need for Sequence and Acknowledgment fields. Similarly, since UDP does not provide a flow control mechanism, the TCP Window field is removed. The result of UDP performing a best effort delivery mechanism is a relatively small transport layer protocol header, with the protocol relatively simple in

Source Port	Destination Port
Message Length	Checksum

Exhibit 21. The User Datagram Protocol Header

comparison to TCP. Because the best way to understand the operation of UDP is via an examination of its header, let us do so. Before we do, as a reminder note that similar to TCP, an IP header will prefix the UDP header, with the resulting message consisting of the IP header, UDP header, and user data referred to as a UDP datagram.

- *Source and Destination Port fields* — The Source and Destination Port fields are each 16 bits or 2 bytes in length and function in a manner similar to their counterparts in the TCP header. That is, the Source Port field is optionally used, with a value either randomly selected or filled in with zeroes when not in use, while the destination port contains a numeric that identifies the destination application or process.
- *Length field* — The Length field indicates the length of the UDP datagram to include header and user data that follows the header. This 2-byte field has a minimum value of 8 that represents a UDP header without data.
- *Checksum field* — The Checksum field is 2 bytes in length. The use of this field is optional and its value is set to 0 if the application does not require a checksum. If a Checksum is required, it is calculated on what is referred to as a pseudo header. The pseudo header is a logically formed header that consists of the source and destination addresses and the protocol field from the IP header. By verifying the contents of the two Address fields through its Checksum computation, the pseudo header assures that the UDP datagram is delivered to the correct destination network and host on the network. This does not verify the contents of the datagram.

Operation

Because the UDP header does not include within the protocol an acknowledgment capability or a sequence numbering capability, it is up to the application layer to provide this capability. This enables some applications to add this capability whereas other applications that run on top of UDP may elect not to include one or both. As previously described, a UDP header and its data are prefixed with an IP header to form a data frame. Upon receipt of the datagram, the IP layer strips off that header and submits the remainder to UDP software at the transport layer. The UDP layer reads the destination port number as a mechanism to de-multiplex the data and send it to its appropriate application.

Applications

The UDP protocol is primarily used by applications that transmit relatively short segments and for which the use of TCP would result in a high level of overhead in comparison to UDP. Common examples of applications that use UDP as a transport protocol include the Simple Network Management Protocol (SNMP), Domain Name Service (DNS), and the newly emerging series of applications from numerous vendors that transport digitized voice over the

Internet and are collectively referred to as Internet Telephony. Concerning the latter, most implementations of Internet Telephony applications use both TCP and UDP. TCP is used for call setup, whereas UDP is used to transport digitized voice once the setup operation is completed. Because real-time voice cannot tolerate more than a fraction of a second of delay, Internet applications do not implement error detection and correction, because retransmissions would add delays that would make reconstructed voice sound awkward. Instead, because voice does not rapidly change, applications may either "smooth" an error or drop the datagram and generate a small period of noise that cannot effect the human ear. This is because most Internet Telephony applications transmit 10-ms or 20-ms slices of digitized voice, making the error or even the loss of one of a few datagrams transmitting such slides of a conversation most difficult to notice.

Chapter 3

IPv4 Addressing Basics and the DNS

In Chapter 2, we examined the TCP/IP protocol suite. We noted that the 32-bit IP address used in the IP header defines the location of the destination address of a datagram as well as the originator of the datagram, with the latter referred to as the source IP address.

In the wonderful world of IP addressing, there are actually two types of IP addresses: one that uses 32 bits and another that uses 128 bits. The first type of IP address is supported by version 4 of the IP protocol, referred to as IPv4, and is the primary focus of this chapter. The second type of IP addressing, referred to as IPv6 because it is supported by a new version of the IP protocol, will be discussed in Chapter 8.

This chapter will focus attention on IPv4 addressing basics and the Domain Name System (DNS). IP addressing forms the basis for configuring workstations and servers, routers, and gateways, as well as DNS servers, and provides the mechanism that ensures datagrams reach their intended destination. DNS servers that form the DNS provide a mechanism to translate host names in the form of English or near-English mnemonics into IP addresses necessary for the actual routing of data. Thus, the two key topics covered in Chapter 3 can be viewed as providing a mechanism for the routing of information on an intranet and Internet basis.

IPv4 Addressing

IP addresses are used by the IP to identify distinct devices, such as hosts, routers, and gateways, as well as to route data to those devices. Each device in an IP network must be assigned to a unique IP address so that it can receive communications addressed to it.

IPv4 uses 32-bit binary numbers to identify the source and destination addresses in each packet. This address space provides 2,294,967,296 distinct addressable devices, which exceeded the population of the world when the Internet was initially developed. However, the proliferation of personal computers, the projected growth in the use of cable modems which require individual IP addresses, and the fact that every interface on a gateway or router must have a distinct IP address has contributed to a rapid depletion of available IP addresses. Recognizing that hundreds of millions of people in densely populated countries such as China and India may eventually be connected to the Internet, and also recognizing the potential for cell phones and even pacemakers to communicate via the Internet, the Internet Activities Board (IAB) commenced work on a replacement for the current version of IP during 1992. Although the addressing limitations of IPv4 were of primary concern, the efforts of the IAB resulted in a new protocol with a number of significant improvements over IPv4 to include the use of 128-bit addresses for source and destination devices. This new version of IP, which is referred to as IPv6, was finalized during 1995 and is currently being evaluated on an experimental portion of the Internet. Because this chapter is concerned with IPv4 addressing, we will cover the addressing schemes, address notation, host address restrictions, and special addresses associated with IPv6 in Chapter 8.

Working with Binary and Hexadecimal Numbering Systems

Both IPv4 and IPv6 addressing represent the use of binary numbers. Because it is often convenient to represent binary numbers in hexadecimal numbers, we will briefly examine both numbering systems. This will serve as both a review as well as to make us appreciate the use of dotted decimal notation described later in this chapter.

If we are familiar with the decimal numbering system, it becomes a relatively easy process to understand binary. However, instead of thinking of a decimal number as *a number*, we need to examine the number by the positions of each digit to obtain an appreciation for the binary numbering system. For example, consider the number 237 in decimal. That number represents two hundreds (2×10^2) plus three tens (3×10^1) plus seven ones (7×10^0). Thus, on a digit positional basis, we could rewrite the number 237 as:

$$2 \times 10^2 + 3 \times 10^1 + 7 \times 10^0$$

remembering that any number to the zero power has a value of one.

Similar to the decimal system, the binary system uses positional notation. However, each position in the binary numbering system only has two possible values instead of ten. Also, each position in the binary numbering system represents a power of two instead of a power of ten. Thus, the value of an 8-bit binary number becomes:

$$n \times 2^7 + n \times 2^6 + n \times 2^5 + n \times 2^4 + n \times 2^3 + n \times 2^2 + n \times 2^1 + n \times 2^0$$

where n is either 0 or 1. For example, the number 10111010 in binary, commonly noted as 10111010_2, can be converted to its decimal equivalent by computing the positional value of each bit position as follows:

$$1 \times 2^7 + 0 \times 2^6 + 1 \times 2^5 + 1 \times 2^4 + 1 \times 2^3 + 0 \times 2^2 + 1 \times 2^1 + 0 \times 2^0 = 186$$

As indicated in the preceding example, the weight of each position in a binary number is two times the weight of the one to its immediate right. To facilitate the conversion process between binary and decimal numbers, Exhibit 1 contains a table of powers of two up to 2^8, their positional weight, and their equivalent binary number.

In the hexadecimal numbering system, we use a base of 16. Thus, a number in the hexadecimal system can be expressed as a series of multipliers of 16 to a power. For example, a four-digit hexadecimal number could be expressed as:

$$n \times 16^3 + n \times 16^2 + n \times 16^1 + n \times 16^0$$

However, unlike the binary system where n is either 0 or 1, in the hexadecimal numbering system each digit can range in value from 0 to 15; however, the digits 10 through 15 are represented by the letters A through F to avoid confusion with the decimal numbering system.

To convert a hexadecimal number to decimal, we need to multiply each digit by the value of the base and add the results. For example, consider the hexadecimal number 3A10, which is normally referenced as h3A10 to indicate hexadecimal representation. That number is equivalent to:

$$3 \times 16^3 + 10 \times 16^2 + 1 \times 16^1 + 0 \times 16^0$$

$$\text{or } 12288 + 2560 + 16 + 0 = 1484_{10}$$

Exhibit 1. Powers of Two

Power of Two	Positional Weight	Binary Number
2^0	1	0
2^1	2	10
2^2	4	100
2^3	8	1000
2^4	16	10000
2^5	32	100000
2^6	64	1000000
2^7	128	10000000
2^8	256	100000000

Exhibit 2 illustrates the relationship between decimal, hexadecimal, and binary for one-digit position in a hex number.

Exhibit 2. Relationship between Decimal, Hexadecimal, and Binary for One-Digit Position in a Hex Number

Decimal	Hexadecimal	Binary
0	0	0000
1	1	0001
2	2	0010
3	3	0011
4	4	0100
5	5	0101
6	6	0110
7	7	0111
8	8	1000
9	9	1001
10	A	1010
11	B	1011
12	C	1100
13	D	1101
14	E	1110
15	F	1111

If you focus your attention on the relationship between hexadecimal and binary, you will note one of the reasons why hex is popularly used in computer science. That reason is that each hex digit represents four binary digits. Thus, the use of hex provides a shorthand method of writing binary bytes or octets of data. For example, consider the 32-bit number shown below:

01110010111001010110110000100111

This number could be rewritten in hex as h72E56C27. Because it is easy to make an error when entering a long string of digits in any numbering system, a common method to reduce errors is to place a decimal point between certain groups of digits. As we examine the use of dotted decimal notation later in this chapter and the use of hexadecimal notation for entering 128-bit IPv6 addresses in Chapter 8, we will also note the placement of decimal points used to both facilitate the entry of numbers as well as to minimize the potential of errors.

IPv4 Addressing

The IP was officially standardized in September 1981. Included in the standard was a requirement for each host connected to an IP-based network to be assigned a unique, 32-bit address value for each network connection. This requirement resulted in some networking devices, such as routers and gateways, that have interfaces to more than one network, as well as host computers with multiple connections to the same or different network being assigned a unique IP address for each network interface. Exhibit 3 illustrates two bus-based Ethernet LANs connected by the use of a pair of routers. Note that each router has two interfaces, one represented by a connection to a LAN and the second represented by a connection to a serial interface that provides router to router connectivity via a wide area network. Thus, each router will have two IP addresses, one assigned to its LAN interface and the other assigned to its serial interface. By assigning addresses to each specific device interface, this method of addressing enables packets to be correctly routed when a device has two or more network connections.

When the IP was developed, it was recognized that hosts would be connected to different networks and that those networks would be interconnected to form an Internet. Thus, in developing the IP addressing scheme, it was also recognized that a mechanism would be required to identify a network as well as a host connected to a network. This recognition resulted in the development of a two-level addressing hierarchy, which is illustrated in Exhibit 4.

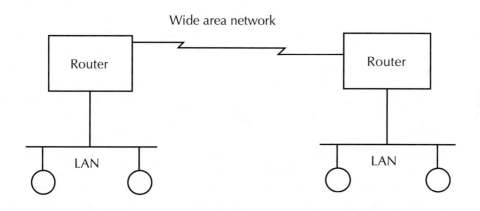

Exhibit 3. Distinct Address Assigned to Each Device Interface in an IP Environment

Network-Prefix	Host-Address

Most IP addresses include a two-level hierarchy consisting of a network prefix and a host address.

Exhibit 4. The Two-Level IP Addressing Structure

Under the two-level IP addressing scheme, all hosts on the same network must be assigned the same network prefix, but must have a unique host address to differentiate one host from another. Similarly, two hosts on different networks must be assigned different network prefixes; however, the hosts can have the same host address.

Address Classes

When IP was standardized, it was recognized that the use of a single method of subdivision of the 32-bit address into network and host portions would be wasteful with respect to the assignment of addresses. For example, if all addresses were split evenly, resulting in 16 bits for a network number and 16 bits for a host number, the result would allow a maximum of 65,532 ($2^{16} - 2$) networks with up to 65,534 hosts per network. Then, the assignment of a network number to an organization that only had 100 computers would result in a waste of 65,434 host addresses that could not be assigned to another organization. Recognizing this problem, the designers of IP decided to subdivide the 32-bit address space into different address classes, resulting in five address classes being defined. Those classes are referred to as Class A through Class E.

Class A addresses are for very large networks, while Class B and Class C addresses are for medium-sized and small-sized networks, respectively. Class A, B, and C addresses incorporate the two-level IP addressing structure previously illustrated in Exhibit 4. Class D addresses are used for IP multicasting, where a single message is distributed to a group of hosts dispersed across a network. Class E addresses are reserved for experimental use. Both Class D and Class E addresses do not incorporate the two-level IP addressing structure used by Class A through Class C addresses.

Exhibit 5 illustrates the five IP address formats to include the bit allocation of each 32-bit address class. In examining Exhibit 5, note that the address class can easily be determined through the examination of the values of 1 or more of the first 4 bits in the 32-bit address. Once an address class is identified, the subdivision of the remainder of the address into the network and host address portions is automatically noted. To obtain an appreciation for the use of each address class, let us examine the composition of the network and host portion of each address when applicable to provide some basic information that can be used to indicate how such addresses are used. Concerning the allocation of IP addresses, it should be noted that specific class addresses are assigned indirectly by the Internet Corporation for Assigned Names and Numbers (ICANN). Typically ICANN assigns blocks of IP addresses to Internet Service Providers (ISPs), with the latter assigning IP addresses to subscribers.

Class A

A Class A IP address is defined by a 0-bit value in the high-order bit position of the address. This class of addresses uses 7 bits for the network portion

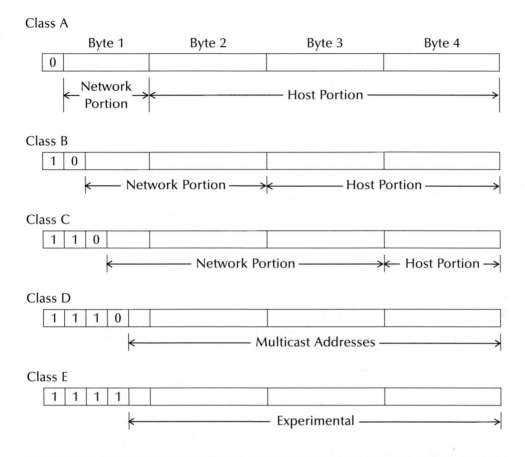

Exhibit 5. IP Address Formats

and 24 bits for the host portion of the address. As a result of this subdivision, 128 networks can be defined with approximately 16.78 million hosts capable of being addressed on each network. Due to the relatively small number of Class A networks that can be defined and the large number of hosts that can be supported per network, Class A addresses are commonly assigned to large organizations and countries that have national networks.

When a Class A address is assigned to an organization, the registration authority assigns a fixed value to the first octet of the address. The last three octets of the Class A address are then managed by the organization assigned the address.

Class B

A Class B network is defined by the setting of the 2 high-order bits of an IP address to 10. The network portion of a Class B address is 14 bits in width, while the host portion is 16 bits wide. This results in the ability of Class B addresses to be assigned to 16,384 networks, with each network having the ability to support up to 65,536 hosts. Due to the manner by which

Class B addresses are subdivided into network and host portions, such addresses are normally assigned to relatively large organizations with tens of thousands of employees.

The assignment of a Class B address results in the registration authority providing a fixed value to the first 2 octets of a 32-bit address. The last two octets are then managed by the organization assigned the Class B address.

Class C

A Class C address is identified by the first 3 bits in the IP address being set to the value 110. This results in the network portion of the address having 21 bits, while the host portion of the address is limited to 8-bit positions.

The use of 21 bits for a network address enables approximately 2 million distinct networks to be supported by the Class C address class. Because 8 bits are used for the host portion of a Class C address, this means that each Class C address can theoretically support up to 256 hosts. Due to the subdivision of network and host portions of Class C addresses, they are primarily assigned for use by relatively small networks, such as an organizational LAN. Because it is quite common for academic institutions, government agencies, and commercial organizations to have multiple LANs, it is also quite common for multiple Class C addresses to be assigned to organizations that require more addresses, but are not large enough to justify a Class B address. Similar to Class A and Class B addresses, the assignment of a Class C address results in a fixed value to a portion of the address while the remainder is assignable by the receiving organization. In the case of a Class C address, the registration authority assigns a fixed value to the first three octets, resulting in the receiving organization controlling the last octet. Although Class A through C addresses are commonly assigned by ICANN to Internet Service Providers for distribution to their customers, Class D and E address classes represent special types of IP addresses.

Class D

A Class D IP address is defined by the assignment of the value 1110 to the first 4 bits in the address. The remaining bits are used to form what is referred to as a multicast address. Thus, the 28 bits used for that address enables approximately 268 million possible multicast addresses.

Multicast is an addressing technique which allows a source to send a single copy of a frame to a specific group through the use of a multicast address. Through a membership registration process hosts can dynamically enroll in multicast groups. Thus, the use of a Class D address enables up to 268 multicast sessions to simultaneously occur throughout the world.

Until recently the use of multicast addresses was relatively limited; however, its use can be expected to considerably increase because it provides a mechanism to conserve bandwidth, which is becoming a precious commodity.

To understand how Class D addressing conserves bandwidth, consider a digitized video presentation routed from the Internet onto a private network for which users working at five hosts on the network wish to review. Without a multicast transmission capability, five separate video streams would be transmitted onto the private network, with each video stream consisting of frames with five distinct host destination addresses. In comparison, through the use of a multicast address, one video stream would be routed to the private network.

Because a video stream requires a relatively high amount of bandwidth in comparison to interactive query-response client-server communications, the ability to eliminate multiple video streams via multicast transmission can prevent networks from being saturated. This capability can also result in the avoidance of session time-outs when client-server sessions are delayed due to high LAN utilization levels, providing another reason for the use of multicast transmission.

Class E

The fifth address class defined by the IP address specification is a reserved address class known as Class E. A Class E address is defined by the first 4 bits in the 32-bit IP address having the value of 1111. This results in the remaining 28 bits being capable of supporting approximately 268.4 million addresses. Class E addresses are restricted for experimentation.

Dotted-Decimal Notation

Recognizing that the direct use of 32-bit binary addresses is both cumbersome and unwieldy to deal with, a technique more user friendly was developed. That technique is referred to as dotted-decimal notation in recognition of the fact that the technique developed to express IP addresses occurs via the use of four decimal numbers separated from one another by decimal points.

Dotted-decimal notation divides the 32-bit IP address into four 8-bit (1-byte) fields, with the value of each field specified as a decimal number. That number can range from 0 to 255 in bytes 2, 3, and 4. In the first byte of an IP address, the setting of the first 4 bits in the byte used to denote the address class limits the range of decimal values that can be assigned to that byte. For example, from Exhibit 5, a Class A address is defined by the setting of the first bit position in the first byte to 0. Thus, the maximum value of the first byte in a Class A address is 127. Exhibit 6 summarizes the number ranges for Class A through Class C IP addresses.

To illustrate the formation of a dotted-decimal number, first focus attention on the decimal relationship of the bit positions in a byte. Exhibit 7 indicates the decimal values of the bit positions within an 8-bit byte. Note that the decimal value of each bit position corresponds to 2^n, where n is the bit position in the byte. Using the decimal values of the bit positions shown in Exhibit 7, let us assume the first byte in an IP address has its bit positions set to 11000000.

Exhibit 6.   Class A through C Address Characteristics

Class	Length of Network Address (bits)	First Number Range (decimal)
A	B	0–127
B	16	128–191
C	24	192–223

128	64	32	16	8	4	2	1

The decimal value of the bit positions in a byte correspond to 2^n where n is the bit position that ranges from 0 to 7.

Exhibit 7.   Decimal Values of Bit Positions in a Byte

Then, the value of that byte expressed as a decimal number becomes 128 + 64 or 192. Now assume that the second byte in the IP address has the bit values 01001000. From Exhibit 7 the decimal value of that binary byte is 64 + 8 or 72. Further assume that the last 2 bytes in the IP address have the bit values 00101110 and 10000010. Then, the third byte would have the decimal value 32 + 8 + 4 + 2 or 46, while the last byte would have the decimal value 128 + 2 or 130.

Based on the preceding, the dotted-decimal number 192.72.46.130 is equivalent to the binary number 11000000010010000010111010000010. Obviously, it is easier to work with as well as remember 4 decimal numbers separated by dots than a string of 32 bits.

Reserved Addresses

There are three blocks of IP addresses that were originally reserved for networks that would not be connected to the Internet. Those address blocks were defined in RFC 1918, Address Allocation for Private Internets, and one is summarized in Exhibit 8.

Exhibit 8.   Reserved IP Addresses for Private Internet Use

Address Blocks
10.0.0.0–10.255.255.255
172.16.0.0–172.31.255.255
192.168.0.0–192.168.255.255

Both security considerations as well as difficulty in obtaining large blocks of IP addresses resulted in many organizations using some of the addresses listed in Exhibit 8 while connecting their network to the Internet. Because the use of any private Internet address by two or more organizations connected to the Internet would result in addressing conflicts and the unreliable delivery of information, those addresses are not directly used. Instead, organizations commonly install a proxy firewall that provides address translation between a large number of private Internet addresses used on the internal network and a smaller number of assigned IP addresses. Not only does this technique allow organizations to connect large internal networks to the Internet without needing to obtain relatively scarce Class A or Class B addresses, but in addition the proxy firewall hides internal addresses from the Internet community. This provides a degree of security because any hacker who attempts to attack a host on your network actually has to attack your organization's proxy firewall.

In addition to proxy firewalls providing a network address translation capability, many routers provide this feature. In Chapter 4 when we describe and discuss the role of special IP addresses, we will examine various network address translation methods.

Networking Basics

As previously noted, each network has a distinct network prefix and each host on a network has a distinct host address. When two networks are interconnected by the use of a router, each router port is assigned an IP address that reflects the network to which it is connected. Exhibit 9 illustrates the connection of two networks via a router, indicating possible address assignments. Note that the first decimal number (192) of the 4-byte dotted-decimal numbers associated with two hosts on the network on the left portion of Exhibit 9 denotes a Class C address. This is because 192 decimal is equivalent to 11000000 binary. Since the first 2 bits are set to the bit value 11, this indicates from Exhibit 5 a Class C address. Also note that the first 3 bytes of a Class C address indicate the network while the fourth byte of a Class C address indicates the host address. Thus, the network shown in the left portion of Exhibit 9 is denoted as 192.78.46.0, with device addresses that can range from 192.78.46.0 through 192.78.46.255.

In the lower right portion of Exhibit 9, two hosts are shown connected to another network. Note that the first byte for the 4-byte dotted decimal number assigned to each host and the router port is decimal 226, which is equivalent to binary 11100010. Because the first 2 bits in the first byte are again set to 11, the second network also represents the use of a Class C address. Thus, the network address is 226.42.78.0, with device addresses on the network ranging from 226.42.78.0 to 226.42.78.255.

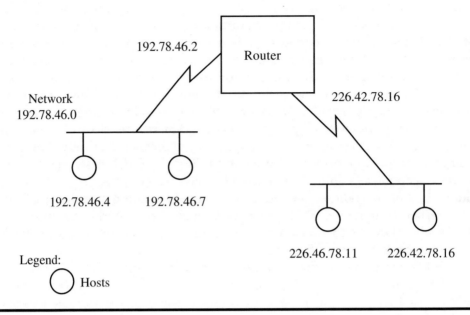

Exhibit 9. IP Address Required for Each Connection in Connecting Routers to Networks

Host Restrictions

Although it would appear that 256 devices could be supported on a Class C network (0 through 255 used for the host address), in actuality the host portion field of an IP address has two restrictions. First, the host portion field cannot be set to all zero bits. This is because an all-0s host number is used to identify a base network. Secondly, an all-1s host number represents the broadcast address for a network or subnetwork. Due to these restrictions, a maximum of 254 devices can be defined for use on a Class C network. Similarly, other network classes have the previously discussed addressing restrictions which reduce the number of distinct addressable devices that can be connected to each type of IP network by two. Since, as previously explained, an all-0s host number identifies a base network, the two networks shown in Exhibit 9 are shown numbered as 192.78.46.0 and 226.42.78.0.

Now that we have an appreciation for basic IP addressing, this chapter will conclude by attention to the manner by which we can enter host names when we want to access a Web or FTP server, Telnet to a certain location, or perform a similar operation without having to remember different IP addresses. The system that provides this translation capability between host names and IP addresses is the Domain Name System (DNS). Although DNS can be viewed as the great facilitator, it is important to remember that all hosts are configured with an IP address. Thus, once we examine the role of the DNS, our discussion of IP addressing will continue in Chapter 4 by noting the role of special IP addresses and it will continue in Chapter 5 with an investigation of subnetting.

The Domain Name System

Although the use of dotted-decimal notation is certainly easier to employ than 32-bit binary addresses, it is still difficult to remember a sequence of four numerics separated by decimal points. This is certainly true when a user requires connectivity with a large number of hosts. In addition, the use of dotted-decimal notation does not provide an indication of the use of the desired host nor its organization. Recognizing these limitations, ARPANET incorporated a mechanism which allowed English-type mnemonics and names to be assigned to hosts.

During the 1970s, when ARPANET provided connectivity for a small number of computers, a single file named HOST.TXT was used to provide a host name to network address translation for each host on the network. That file was maintained at the ARPANET Network Information Center (NIC), and as the network grew in size and complexity, the use of a centrally located file for name to address translations resulted in a series of problems that were eventually alleviated by the implementation of the Domain Name System (DNS). Two of the major problems associated with the use of a single HOST.TXT file included the traffic directed to the host maintaining the file and available name assignments. Concerning network traffic, as more and more hosts were added to the network, traffic routed to the host containing the file HOST.TXT literally exploded, resulting in both transmission and host processing delays. Secondly, the use of a single file for name to address translations precluded two hosts from having the same name, even if they were requested by different organizations. Recognizing these and other problems associated with using a single name to address translation point resulted in the development of a distributed database to perform the same process. That database is the key to the operation of DNS.

Overview

DNS represents a client server approach that was developed to translate English-type names assigned to hosts to their IP addresses. Server, more formally referred to as name server, programs contain information about portions of the DNS database. Information contained in name servers is requested via clients, which are more formally referred to as resolvers. Resolvers are commonly implemented as library routines within an application program, such as FTP. Such routines create queries that are transmitted to a name server. Thus, the entry of an English-type name, such as ftp.gocart.com, as the destination address in an FTP connection would result in the client application sending a request to a name server for the IP address of the destination. That IP address would then represent the address associated with the host name ftp.gocart.com; however, in actuality the host name is ftp and the domain name is gocart.com and the full name ftp.gocart.com is referred to as a fully qualified domain name.

Database Structure

The DNS database is structured similar to an inverted tree, with its root at the top of the structure. Directly under the root are primary domains, with each domain further divided into partitions referred to as subdomains. Exhibit 10 indicates some of the major domain name suffixes assigned by the InterNIC to include a description of the type of organization that is assigned to each suffix.

Readers should note that the entries in Exhibit 10 indicate initially defined generic top level domains. In addition to domains based on organizational category, top level domains are also defined based on the use of two-letter country codes from the International Organization for Standardization's (ISO) 3166 standard. For example, AU identifies the top-level Australian domain, while FR identifies the top-level French domain. Although Exhibit 10 and the two preceding country domain examples use capital letters as domain identifiers, in practice both generic organizational and country top-level domains are specified as dots followed by lower case letters, such as .com, .edu, and .fr. Shortly we will obtain an understanding for the use of the dot as a prefix to the top level generic and country code identifiers.

In addition to the top-level domain names listed in Exhibit 10, during the year 2000 the International Ad Hoc Committee (IAHC), a coalition of international groups whose actions represent a test in Internet governance by consensus, recommended seven new domain names. Exhibit 11 lists the new domain name suffixes IAHC proposed.

While the IAHC recommendations have considerable influence, it is ICANN that has the actual authority for administering domain names. During 2001 ICANN proposed seven new domain name suffixes. Those new domain name suffixes are listed in Exhibit 12. It is entirely possible that by the time you read this book that the domain name suffixes listed in Exhibit 12 will be a reality.

Exhibit 13 provides a general schematic of the DNS database indicating the relationship of six top level generic domains in the DNS to each other as well as to possible subdomains. In addition, the lower left portion of the database indicates a possible registered subdomain managed by the commercial company International GoCart Corporation. The address of each application is formed by adding together the mnemonics assigned to each node upward to the top-level domain they reside in, separating each name from the other by a dot. Thus, what is known as the fully qualified domain name for an ftp server would become ftp.gocart.com.

In examining Exhibit 13, it should be noted that each node represents a portion of a database and the number of domains that can be nested under one another is essentially unlimited. Similar to a file system directory structure, each domain can be identified both relative and absolute to its position in the domain. When a domain is identified with respect to its parent domain, relative addressing is used. In comparison, when a sequence of labels is used to identify a domain with respect to the root of the database, absolute addressing is employed.

To facilitate the use of domain names, each organization registers its name. The registration process results in a domain name consisting of the organization's

Exhibit 10.   Domain Name Suffixes

Suffix	Type of Organization
COM	Commercial organization
EDU	Educational organization
GOV	Government agency
MIL	Military organization
NET	Networking organization
ORG	Not for profit organization
INT	International organization

Exhibit 11.   IAHC Proposed Top-Level Domain Names

Domain	Area of Emphasis
.arts	Culture/entertainment
.firm	Business organization
.info	Information services
.nom	Individual or personal nomenclature
.rec	Recreation/entertainment
.store	Goods for purchase
.web	World Wide Web related activities

Exhibit 12.   ICANN New Domain Name Suffixes

Domain Suffix	Area of Emphasis
.biz	Business organization
.info	Information services
.name	Individual or personal nomenclature
.pro	Professional organization
.museum	Cultural
.aero	Aerospace
.coop	Cooperative organization

name and its domain assignment being registered. For example, returning to our International GoCart Corporation example, assume that company is a commercial organization and registered under the name GoCart. Then, it would be assigned the registered domain name gocart.com. Once an organization has a registered domain name, it can prefix that name to indicate specific hosts or applications residing on a host. For example, "www.gocart.com" and "ftp.gocart.com" could represent a World Wide Web (WWW) server and a File Transfer Protocol (FTP) server. They could also represent two applications residing on a common server.

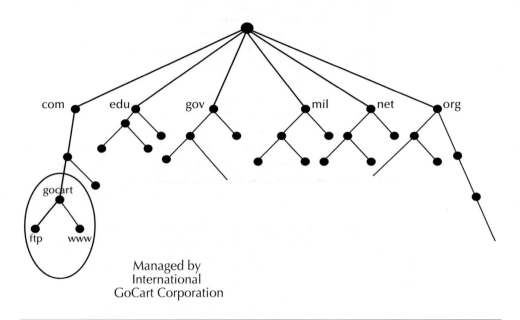

Exhibit 13.   The DNS Database Structure

Name to Address Translation

When you enter the destination IP address in an application as a name, a translation process is required to convert that name into a 32-bit IP address. To accomplish this translation process, each TCP/IP network has a server that keeps track of the host names on the network. When a network user enters a name instead of an IP address, the application's name resolver, functioning as a client, transmits a request to the local name server. If the name resides on the network or was recently obtained from another network and is in cache memory, the server returns the IP address associated with the name. If the name is not in the name server nor in its cache memory, the server will send a request to a "higher" DNS server on another network. In effect, one DNS server has a pointer to another server up the inverted tree structure previously illustrated in Exhibit 13. This forwarding effect can be replicated several times and traverse around the globe until the IP address associated with the English-type name is found and returned. To facilitate the location of names, each server has pointers to servers in other domains. This alleviates, for example, the necessity for searching through several servers in the .edu domain to obtain the IP address for a name registered in the .gov domain.

One of the key advantages of the hierarchical database structure of DNS is its ability to support near duplicate names, enabling, for example, the International Pretzel University to set up a WWW server using the address www.ipretzel.edu. Similarly, a women's clothing store named White House located in the mall at Shelter Cove on Hilton Head Island could set up a WWW server using the address www.whitehouse.com and the NDS process of name to address translation would allow appropriate requests to obtain the IP address of that commercial entity separate from the government entity. In concluding this overview of DNS, it should be noted that the near-English mnemonics and names used to represent distinct host addresses play no role in the actual routing of IP packets. Instead, the DNS process translates those names into IP addresses, which are then used by routers to establish a connection to the desired destination initially defined using near-English mnemonics and names.

Exhibit 14 illustrates an example of a host name to IP address resolution process. In this example, a workstation whose IP address is 205.131.175.6 enters the host name www.whatmeworry.com into the horizontal bar location in a browser where uniform resource locators are placed to indicate an applicable Web page to visit. The TCP/IP protocol stack operating on the user's workstation first checks its cache to determine if the host name's IP address was previously learned. If so, the previously learned IP address is used to request the Web page for the entered URL www.whatmeworry.com. If the address was not previously learned, the workstation will send a request for a DNS lookup, which represents the process whereby an applicable IP address is provided for a fully qualified domain name. One question you might have is how the workstation knows where to send the request for the DNS lookup. The answer to this question is in the configuration of TCP/IP on a workstation.

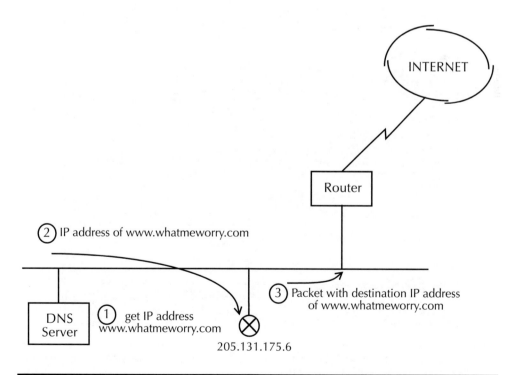

Exhibit 14.   Host Name to IP Address Resolution Process

Exhibit 15 illustrates the settings on the IP (TCP/IP) Properties dialog box under Windows 2000 on the author's computer. In this example, note the lower portion of the dialog box, which enables both a primary (preferred) and alternate DNS server to be defined. In this example the preferred DNS server is located on the same network as the author's computer, while the alternate DNS server is located on a different network.

If you click on the button labeled *Advanced* shown in Exhibit 15, Windows 2000 will display a dialog box labeled Advanced TCP/IP Settings. An example of this dialog box is shown in Exhibit 16. If you examine the top portion of Exhibit 15, you will note that the DNS server addresses are listed in their order of use, with the first entry corresponding to the preferred DNS server address entered in Exhibit 15. Also note the check mark in the lower portion of Exhibit 16. Through this setting, this computer's address will be registered in DNS.

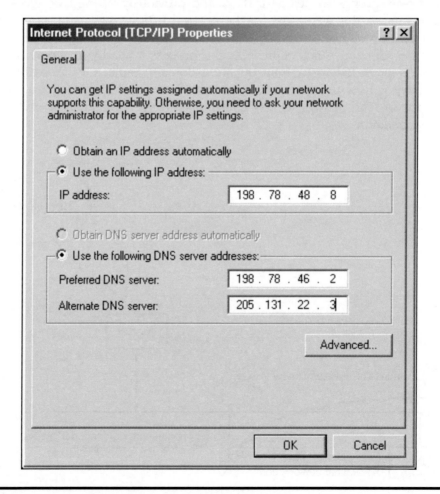

Exhibit 15.   Specifying the Location of Both a Primary (Preferred) and Alternate DNS Server in Windows 2000

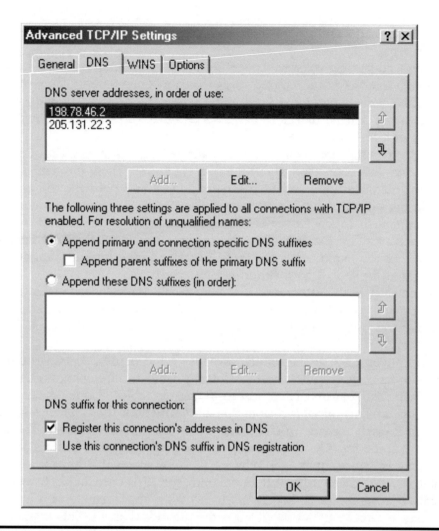

Exhibit 16. Registering an Address in DNS through the Advanced TCP/IP Settings Dialog Box

Inverse Addressing

The previously described DNS lookup operation results in the determination of an IP address based on providing a host name. A second function provided by DNS permits a reserve process to be supported, i.e., you can provide an IP address to DNS to obtain a host name. While this may appear questionable because routing occurs via IP addresses, the ability to perform inverse addressing can be an important consideration when a Webmaster is attempting to identify visitors to their site. After all, would not a report stating that 57 percent of visitors arrived from educational domains (.EDU) be more valuable than a list of IP addresses? Through the inverse addressing process, it becomes possible to check DNS table entries and determine host names associated with IP addresses. Once this is accomplished it then becomes possible to note the domains of visitors and gather more meaningful statistics than a simple list of IP addresses.

Inverse addressing operates based on tables maintained on the previously mentioned Internet hosts referred to as Domain Name Servers. As previously noted, DNS servers provide a local name service that maps host names to IP addresses and domain names to name servers that know about the collection of hosts in a domain.

To illustrate the operation of inverse addressing, assume GoCart Corporation's network is connected to the Internet via an ISP. Then, the names of the publicly accessible hosts and Web servers on GoCart's network will be registered in the ISP's Domain Name Server. The ISP's DNS is then designated as the authoritative Name Server for the gocart.com domain as well as other domains served by the ISP. This also means that the ISP Name Server is registered as the name server for those domains in the Internet's master root zone files at the top of an inverted tree hierarchy.

In actuality, each ISP, unless it is very small, typically maintains two or more Name Servers. This provides a level of redundancy and the root zone file at the top of the hierarchy would then contain the IP addresses of each Name Server.

Under inverse addressing which provides a mechanism to translate an IP address to its corresponding host or domain name, a different set of tables is used. Each DNS that maintains host-to-IP address tables also maintains tables for use by .arpa, a special top-level domain. Named for the original Advanced Research Projects Agency (ARPA) from which the Internet evolved, the .arpa domain is at the same level of the top-level domains, such as .com, .edu, .mil, and so on. However, unlike the other top-level domains that have lower-level domains containing near-English mnemonics, the .arpa domain has lower-level domains that represent IP addresses. In actuality, the .arpa domain has only one domain directly under it, which is the .in-addr domain. This domain has a large number of subdomains, each representing an IP address that appears elsewhere in DNS tables as an association with a host name. Actual IP addresses in the subdomains are arranged backward from the normal IP address. For example, if the IP address of a host is 205.135.173.8, it is placed in the .in-addr domain entry as 8.173.135.205. The rationale for the backward placement of IP addresses results from the hierarchical structure of host names and IP addresses.

To illustrate the rationale for the reverse address, let us assume GoCart Corporation's Web server has the IP address 205.135.173.8. When using host names such as www.gocart.com, the lowest level of the naming hierarchy (www) represents the least significant portion of the host name. However, under the hierarchy of numeric IP addressing, the opposite situation occurs, i.e., for the address 205.135.173.8, the most significant portion of the address is the network portion (205) while the least significant portion (8) comes last.

Operation

Let us assume the Web server www.gocart.com wants to determine who is visiting its site. When it receives a hit on its home page, it knows who transmitted the hit since the source IP address is contained in the IP header of the datagram transporting Web traffic. To perform a reverse address lookup, the Web server will transmit a reverse lookup request to its DNS. That request provides the IP address and requests the host name. The DNS will first check its local reverse lookup table to determine if it has an entry. If so, it will return the host name to the Web server. If not, it will send a request to the Name Server for the correct in-addr.arpa domain which will return the desired host name.

Prior to moving on, it should be mentioned that we will return to the settings or configuration of workstations in Chapters 4 and 5. The rationale for not mentioning the specific configuration of a workstation for TCP/IP operations at this time is because we need to learn a bit more, no pun intended, about IP addressing. Thus, after we discuss the role of special IP addresses in Chapter 4 and subnetting in Chapter 5, we will then have the background to understand the do's and don'ts associated with IP addressing and may turn our attention to the configuration process.

Returning to the DNS process, let us assume our request for a DNS lookup to the preferred DNS server does not result in a match for our query. That server has a pointer to a "higher" DNS server, which might be the DNS server of our Internet Service Provider. Thus, the preferred DNS server now sends a query to the ISP DNS server. If the ISP DNS server does not have an entry for the lookup request, it uses its pointer record to send a request to a higher DNS server, perhaps one located at a Network Service Provider (NSP). Each time the request is forwarded, an additional time delay occurs based on network traffic and the activity of each DNS server. Thus, one of two things will happen. Either a timer set to a predefined value on the workstation will expire and an error message will be displayed or the lookup will succeed and the applicable IP address will be returned to the workstation. When the first situation occurs, the workstation will display a dialog box asking you to check the name entered. After you click on the button labeled OK, you can simply try again. There is a very good chance that your second request, if your spelling was in order, to go to a new address will succeed. This is because as DNS requests went up the inverted tree hierarchy, the response flowed down the hierarchy and intermediate DNS servers placed the learned information into memory. Thus, while it took too long to retrieve the required information the first time, when the second request occurred, it did not have to flow up the entire hierarchy. If the second situation occurs, the host name and IP address are entered in the preferred DNS server's cache as well as your workstation's cache memory, simplifying the next request to the host name.

Now that we have an appreciation for IPv4 addressing and the role of the DNS, we will move on to Chapter 4 to learn the role of special IP addresses.

Chapter 4

The Role of Special IP Addresses

In Chapter 3, we focused on understanding basic IPv4 addressing and the role of the Domain Name System. We also reviewed binary and hexadecimal numbering and the composition of different IP addresses. Although we briefly mentioned RFC 1918 addresses in Chapter 3, a discussion of the role of special IP addresses (to include RFC 1918 addresses) was deferred until this chapter. Thus, the purpose of this chapter is to obtain a detailed understanding of the role of special IP addresses and the restrictions they place on the configuration and operation of devices in a TCP/IP network environment.

The Loopback Address

When IPv4 addresses were being classified, it was recognized that there would be a need to test the protocol stack. This need was satisfied through the assignment of a special address referred to as a loopback address.

Address Range

By convention, any address commencing with 127 as its first dotted decimal position is reserved for a loopback address. Unfortunately, this scheme reserves the entire Class A address space for network 127.0.0.0, resulting in 2^{16} numbers being reserved for loopback addresses. In practice, as we will shortly discuss, not all operating systems reserve the entire Class A 127 network for loopback addresses.

Utilization

Exhibit 1 illustrates the use of the Ping command under the Windows command prompt to ping the network address 127.0.0.1. That address is used by all operating systems familiar to this author for loopback. In examining Exhibit 1, note that even though this author's PC was disconnected from his network, pinging the loopback address resulted in four packets being transmitted and returned because they are simply looped through the computer. As you might expect, this action occurs at electronic speed, which explains why the round trip times were recorded as 0 ms at the lower portion of Exhibit 1. For those readers not familiar with the use of Ping, its use will be covered in detail in Chapter 9 when we discuss network utility tools. However, by noting that packets sent were received, we also know that the TCP/IP protocol stack is operating correctly.

Two additional examples of the use of the 127 network address are contained in Exhibit 2. At the top of Exhibit 2, the Ping command was used with the IP address 127.0.0.0. Note that Windows informs us that the *destination specified is invalid* because the address represents the 127 network and not a device on the network. In the second example shown in Exhibit 2, the network address 127.0.0.2 was used. While the Windows 2000 operating system recognizes this address as host 2 on network 127 as a valid loopback address, it is important to note that not all operating systems recognize a loopback address other than 127.0.0.1. Thus, when testing your protocol stack, you typically cannot go wrong by using 127.0.0.1 which is recognized by all operating systems.

```
Command Prompt                                                      _ □ ×
Microsoft Windows 2000 [Version 5.00.2195]
(C) Copyright 1985-1999 Microsoft Corp.

C:\>ping 127.0.0.1

Pinging 127.0.0.1 with 32 bytes of data:

Reply from 127.0.0.1: bytes=32 time<10ms TTL=128
Reply from 127.0.0.1: bytes=32 time<10ms TTL=128
Reply from 127.0.0.1: bytes=32 time<10ms TTL=128
Reply from 127.0.0.1: bytes=32 time<10ms TTL=128

Ping statistics for 127.0.0.1:
    Packets: Sent = 4, Received = 4, Lost = 0 (0% loss),
Approximate round trip times in milli-seconds:
    Minimum = 0ms, Maximum =   0ms, Average =   0ms

C:\>
```

Exhibit 1. Testing the TCP/IP Protocol Stack through a Loopback by Transmitting and Receiving Datagrams

```
[CN] Command Prompt                                                    _ □ X
Pinging 127.0.0.0 with 32 bytes of data:

Destination specified is invalid.
Destination specified is invalid.
Destination specified is invalid.
Destination specified is invalid.

Ping statistics for 127.0.0.0:
    Packets: Sent = 4, Received = 0, Lost = 4 (100% loss),
Approximate round trip times in milli-seconds:
    Minimum = 0ms, Maximum =   0ms, Average =   0ms

C:\>ping 127.0.0.2

Pinging 127.0.0.2 with 32 bytes of data:

Reply from 127.0.0.2: bytes=32 time<10ms TTL=128
Reply from 127.0.0.2: bytes=32 time<10ms TTL=128
Reply from 127.0.0.2: bytes=32 time<10ms TTL=128
Reply from 127.0.0.2: bytes=32 time<10ms TTL=128

Ping statistics for 127.0.0.2:
    Packets: Sent = 4, Received = 4, Lost = 0 (0% loss),
Approximate round trip times in milli-seconds:
```

Exhibit 2. Two Examples of the Ping Command Using Network 127 Class A Addresses

All 0s

The IP address of 0.0.0.0 represents another special address. As we will soon note, although this address was once used fairly commonly, it should not be used today.

Utilization

An all-0s address, referred to as *all zeros*, is used as a default route in a routing table. However, under most router operating systems, the use of this all-zeros address is disabled. Thus, you must enable this address to use it as a default route.

A second use of an all-zeros IP address is as a source address in a boot configuration request. At one time when PCs were expensive, it was common to have diskless workstations that on power-up downloaded a configuration from a server. When this occurred, the device had yet to obtain an IP source address. Thus, it would use the source address of all zeros.

Rationale to Block

As previously noted, it is extremely rare to use an all-0s address today. For this reason, you should block datagrams with a source address of all 0s from entering your network, a function referred to as anti-spoofing which will be discussed at the end of this chapter.

Broadcast Address

A broadcast address represents an address transmitted to every host on a network. In actuality only one datagram is transmitted; however, because of the contents of the destination address in the IP header, each device on the network reads the datagram. There are several IP address patterns that are used for broadcasts. In general, a datagram can be broadcast to one or more networks within a particular network hierarchy.

Basic Broadcast Address

The basic IP broadcast address is 255.255.255.255, which represents an address consisting of all ones or 32 ones under IPv4 addressing. The use of this address results in a datagram being broadcast to each system on the local link.

Utilization

The primary use of an all-ones broadcast address is in the BOOTP and DHCP protocols. Under these protocols, a client does not know its IP address. Thus, it uses the broadcast address of 255.255.255.255 to obtain its IP address and other initialization data from a boot server. For example, a client would transmit a boot request to address 255.255.255.255 and use the reserved address of all zeros (0.0.0.0) as its source IP address.

Use of All Zeros

During the early days in the evolution of the TCP/IP protocol stack, an all-zeros address was used instead of an all-1s address for broadcasting. This practice represents a nonstandard use of an antiquated method of broadcasting and should be avoided.

Broadcasts to Networks

A second version of the broadcast address is to transmit it in a datagram to a specific remote network. This type of broadcast is referred to as a broadcast to a network. As we will soon note, this can be very dangerous.

Utilization

You can transmit an IP datagram to every host on a selected network by setting the entire host portion of the address to all ones. For example, assume you want to transmit a message to every host on the Class C network 205.131.175.0. Because the host portion of the IP address is the fourth byte of the address, you would set the broadcast address as 205.131.175.255.

Exhibit 3 lists the 1s positions for Class A, B, and C broadcast addresses. Note that all 1s only occurs in the host portion of each address because the network portion must remain as is for the datagram to be correctly routed to its destination.

Exhibit 3. IP Address Class Broadcast Addresses

Address Class	Broadcast Address
A	x.255.255.255
B	x.x.255.255
C	x.x.x.255

Note: x indicates a valid dotted decimal number.

Danger in Use

To obtain an appreciation of why the directing of a broadcast to a remote location can be dangerous, consider the network shown in Exhibit 4. In this example, we will assume that the network connected to the Internet has the Class C address 205.131.175.0. This means there can be up to 254 devices connected to that network.

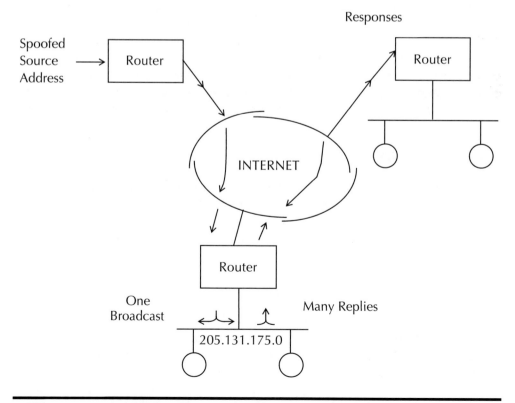

Exhibit 4. Directed Broadcasts Attacking Two Networks

Assume a distant party transmits a Ping to the broadcast address of the network shown in Exhibit 4. To do so, the person simply enters the following command:

Ping 205.131.175.255

This entry results in the transmission of four pings to the router serving the 205.131.175.255 network, where the datagram containing each ping is converted into an Ethernet frame and transmitted as a data link Layer 2 Media Access Control (MAC) broadcast onto the network. Each active powered-up station responds to each ping, resulting in up to 254 responses flowing on the distant network to the router and then over its serial connection to the Internet.

Because each ping command generates four datagrams containing the applicable ICMP message, the remote network could issue up to 254×4 or 1016 responses. Now assume the pingologist decides to use an option that results in the continuous pinging of the distant network's broadcast address. To make matters a bit worse, assume the distant pingologist hides his or her true network address by configuring their workstation with a different IP address than their valid address. Because routers only check the destination address in IP datagrams, this is a perfectly valid method for hiding one's IP address, commonly referred to as address spoofing. To make matters even more alarming, assume the pingologist configured his or her IP address to that of the FBI Web server. This means that a two-phase attack could result. Directed broadcasts to the 205.131.176.255 address first result in a continuous transmission of echo replies generated by devices on that network. Then, because the FBI's Web server address was configured as the pingologist's IP address, the responses now flow toward the FBI server. Thus, instead of being able to generate a traffic load that can adversely affect one network, the person pinging away now has the capability to adversely affect two networks.

Countermeasure

Because of the potential adverse effect from directed broadcasts, most organizations configure their routers to drop directed broadcasts. In a Cisco router environment, you would use the command around no IP-directed broadcasts to preclude directed broadcasts from getting through your router.

Subnet Broadcast Addresses

Although we will defer discussion of subnets and subnetting to Chapter 5, it is important to note that each subnet has a broadcast address. This means it is possible for a broadcast to be directed to a specific subnet. The subnet can be a one that a host is directly connected to or one that is remote from the source host.

If the destination subnet is remote from the source host, the result will be similar to the previously mentioned direct broadcast, i.e., one IP datagram with the broadcast address of a subnet would reach a router connected to the subnet. The router would then transmit a Layer 2 Medical Access Control (MAC) frame with a physical broadcast address in the destination address of the frame. Each host active on the subnet would then respond.

Reserved Address

As noted in Chapter 3, each network cannot use the host address of 0 or 255. A similar restriction concerning the broadcast address of subnets applies, i.e., the broadcast address of a subnet represents a reserved IP address that cannot be assigned to a host. Chapter 5 will examine in detail how to divide a network address into subnets and how to determine the broadcast address on each subnet.

RFC 1918 Addresses

As previously noted in Chapter 3, there are three blocks of addresses that are reserved for use for networks that will not be connected to the Internet. Those address blocks are:

> 10.0.0.0 to 10.255.255.255
> 172.16.0.0 to 172.31.255.255
> 192.168.0.0 to 192.168.255.255

Utilization

Perhaps the most common use of RFC 1918 addresses is as a mechanism to alleviate the shortage of IPv4 addresses, i.e., an organization with thousands of hosts could use RFC 1918 addresses behind a router or firewall that is connected to the Internet. Through the process of Network Address Translation (NAT) (covered in Chapter 7), it becomes possible to translate RFC 1918 addresses into a valid Class C IP address. An example of this method of address translation is shown in Exhibit 5. In examining Exhibit 5, note that the use of a single Class C network address results in 254 valid IPv4 addresses being used to support an entire organization's infrastructure, which could theoretically represent tens of thousands of hosts.

Key Disadvantage

While RFC 1918 addresses provide a significant benefit by extending the use of scarce IPv4 addresses, they also have one key disadvantage. That disadvantage results from the fact that many hackers like to use such addresses as their source address when attacking a network. This use results from the fact that source addresses are not checked by routers. In addition, because RFC

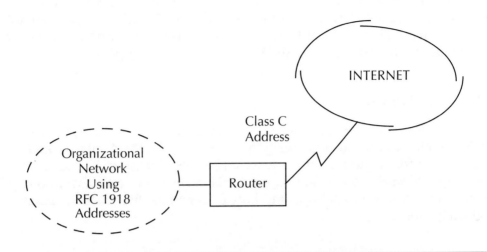

Exhibit 5. One Class C Address Satisfying the Requirements of a Large Organization through Network Address Translation

1918 addresses are not valid Internet addresses, responses to queries that use such addresses in the destination address field will never reach a valid destination. Instead, they will eventually flow into the great bit bucket in the sky as datagrams flow from router to router and the IP hop count field decrements to zero.

Anti-Spoofing Considerations

In concluding this chapter, we will turn to a technique you can consider to minimize the potential for hackers to attack your network using illegitimate or potentially questionable IP source addresses. Exhibit 6 lists applicable statements for a Cisco router that would filter or block RFC 1918 addresses, all zeros, all ones, loopback, and Class D and Class E addresses. Concerning the latter two addresses, they should only be blocked if your network does not support multicasting and the use of experimental IPv4 addresses.

In examining the entries in Exhibit 6, an explanation is in order for those readers not familiar with Cisco access lists. The access list used in Exhibit 6 represents what is referred to as an extended IP access list. The general format of an extended IP access list is:

```
access-list list # {permit 1 deny}[protocol][source
address][wildcard mark][destination address][wildcard
mask][options]
```

Here the list # represents a number between 100 and 199 which informs the router that the access list is an extended IP access list. All statements in one extended access list would use the same list #, which explains why all access list statements shown in Exhibit 6 have the same list number.

Exhibit 6.   Cisco Anti-Spoofing Entries to Consider for a Router's Access List

```
!deny RFC 1918 source addresses
access-list 101 deny ip 192.168.0.0 0.0.255.255 any log
access-list 101 deny ip 10.0.0.0 0.255.255.255 any log
access-list 101 deny ip 172.16.0.0. 0.15.255.255 any log
!deny all zeros, all ones, loopback
access-list 101 deny ip host 0.0.0.0 any log
access-list 101 deny ip host 255.255.255.255 any log
access-list 101 deny ip 127.0.0.0 0.255.255.255 any log
!deny Class D and Class E
access-list 101 deny ip 224.0.0.0 31.255.255.255 any log
access-list 101 deny ip 240.0.0.0 7.255.255.255 any log
```

Each access list statement will contain either a permit or deny keyword. Permit results in the acceptance of a packet that matches the remainder of the statement, while the keyword deny results in the packet matching the criteria in the statement being sent to the great bit bucket in the sky. All of the access list statements contained in Exhibit 6 include a deny keyword because we wish to block the flow of packets that contain source addresses that represent commonly used spoofed addresses. The keyword following deny in each statement is ip, which represents the protocol to be filtered. Because an ip header prefixes TCP, UDP, and even ICMP, this means that all TCP/IP protocols will be examined. Following the protocol are the source address and wildcard mask followed by the destination address and wildcard mask.

Under Cisco terminology, the wildcard mask serves as a mechanism to denote the source or destination hosts on a network to be filtered. A binary 0 is used to indicate a match, while a binary 1 is used to indicate a do not care condition. For an example of the use of the wildcard mask, consider the first access list statement in Exhibit 6 which is employed to block all datagrams that have a source address on the 192.168.0.0 network. Because we need to match the first two octets of the IP address, the first 2 bytes of the wildcard mask are set to 0.0. Next, because the next 2 bytes on the 192.168.0.0 network can be any address, the wildcard mask uses 255.255 in its last two positions. Thus, the source address and source wildcard mask are 192.160.0.0.0.0.255.255. If we continue examination of the first access list statement in Exhibit 6, we will note the keyword *any*. That keyword is equivalent to a source address of 0.0.0.0 and a wildcard mask of 255.255.255.255. Last but not least, because each time a datagram is discarded due to containing a suspicious IP address could represent a potential network attack, we use the log command to record the event.

While there is much more to network security than anti-spoofing statements, their use in a router access list represents a good start toward locking the door.

Chapter 5

Subnetting

Chapter 5 will focus on the subdivision of the two-level hierarchy associated with Class A, B, and C networks into a three-level hierarchy. This subdivision is referred to as subnetting. It provides a considerable degree of IP addressing flexibility for network administrators. In discussing subnetting we will first examine the rationale behind the concept. Once this is accomplished, we will turn to the manner by which subnetting occurs, the use of the subnet mark, how subnetting can be used by organizations, and the manner by which different network devices can be configured to support a three-level hierarchy.

Rationale

Through the mid-1980s, the gradual growth in the use of the Internet resulted in several IP addressing-related problems. Those problems included the waste of address space that could occur when multiple networks within an organization are assigned individual IP addresses, the size of router tables, and the potential management nightmare resulting from attempting to administer a Class A or Class B network as a single entity. As we will shortly note, each of these problems was a driving force for a mechanism to break networks into smaller entities referred to as subnets.

Waste of IP Address Space

One of the problems associated with the use of IP addresses was the original necessity to assign a distinct network address to each network. This method of address assignment could result in the waste of many addresses. Because there are only a finite number of IPv4 addresses, it became recognized that a mechanism to permit a more efficient use of IP addresses was required.

To obtain an appreciation for how the assignment of a Class A, B, or C network address on an individual basis to a network wastes address space, consider Exhibit 1. In this illustration, two Class C networks are shown connected to a router which in turn is connected to the Internet.

In examining Exhibit 1, assume that each Class C network supports 24 workstations and servers. Adding an address for the router interface to each network, each Class C network would use 25 out of 254 available addresses. Thus, the assignment of two Class C addresses to an organization with a requirement to support two networks with a total of 50 devices would result in 458 available IP addresses in effect being wasted. Considering this waste of IP address space on a global basis, it became apparent that assigning individual IP addresses to each network regardless of the number of hosts on a network would rapidly deplete available addresses.

Router Routing Table Size

A second problem associated with the assignment of IP addresses to individual networks concerns the size of router routing tables. As the number of networks proliferated, routers had to learn additional IP addresses in order to route datagrams toward their ultimate destination. Not only did the process of storing additional IP addresses consume additional memory, but, in addition, it required additional time for the router to search its address table entries in an attempt to ascertain onto which port to forward a packet.

Exhibit 2 illustrates an example of the manner by which entries in a router's address table take up storage and require additional processing time. In this

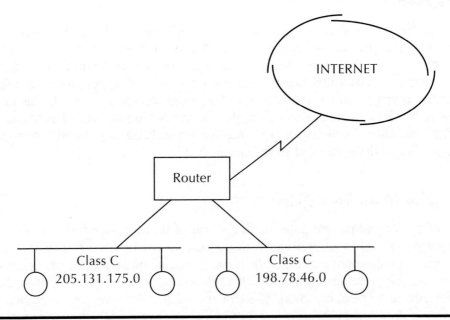

Exhibit 1.   Connecting Two Class C Networks to the Internet

example assume we have a three-port router and a datagram arrives on Port 1. The router examines the destination IP address contained in the IP header and uses that address to search its routing table. That table consists of a sorted list of learned IP addresses and the port associated with the path to the IP address. In actuality this represents a simplified illustration of a routing table because such tables can include alternate paths, a time stamp that governs when old addresses are purged from the table, and other information.

In examining Exhibit 2, assume an IP datagram with the destination network address 198.83.12.0 flows into the router on Port 0. The router will search its routing table to determine the appropriate port onto which to output the datagram.

The right of Exhibit 2 illustrates a small routing table with four network addresses and their associated router ports. For this particular search, only two entries were required to be compared prior to a match occurring between the address in the IP datagram and the corresponding address in the routing table. Now assume the number of routing table entries increased to 10,000. Without a hashing algorithm, the router would, on the average, have to search 5000 entries until a match occurred. Using a simple hash algorithm, the router would start at the middle of the sorted table and determine if the address to be matched was in the upper or lower half of the table. Thus, the first comparison would eliminate half of the table. The second comparison would perform a similar operation, isolating the potential search to a quarter of the table. Thus, five comparisons would quickly isolate the potential linear search to a maximum of 625 table entries. While an extensive amount of research effort went into developing a sophisticated router table search algorithm, a simple fact of life is that the more table entries there are, the longer it will

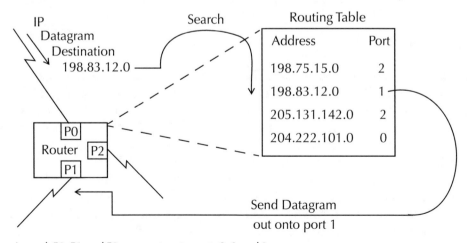

Legend: P0, P1, and P2 represent router ports 0, 1, and 2.

Exhibit 2. More Networks Added to the Internet Increased Size of Routing Tables, Requiring More Memory and Additional Processing Time

take to locate a particular entry. Thus, additional router table entries both consume memory and result in additional processing time.

Management and Performance

A third problem which is primarily associated with using a Class A or Class B address "as is" is management, with a related problem being the performance level of the network. For example, without a subdivision capability, you could have 16,777,214 hosts on a Class A network, while a Class B network could have up to 65,534 hosts. Even if you have a staff of helpful personnel to manage a network that can consist of tens of thousands to tens of millions of hosts, just think about the level of performance of an entity of either of those magnitudes. By permitting the subdivision of large networks into smaller entities, you achieve the ability to either directly manage groups of smaller networks or assign such tasks to departments within an organization that becomes responsible for managing their portion of the network. In addition, because network traffic now flows over many essentially autonomous networks instead of a single network, this subdivision can be expected to relieve congestion by isolating traffic that does not need to be routed to different subnets, i.e., traffic on a subnet destined to a server on that subnet remains on that network and only affects the transmission capability of other stations on that subnet. In comparison, if no subnets existed, then each transmission would affect every workstation on the larger network.

While management and performance issues are normally related to Class A and Class B networks, it is also possible to encounter such problems on a Class C network. Thus, subnetting represents a technique to resolve management and performance problems across all three types of class networks.

RFC 950

Recognizing the previously described problems resulted in RFC 950 being issued in 1985. Entitled Internet Standard Subnetting Procedure, this RFC describes in detail the procedure required to divide or subnet a single Class A, B, or C network into subnetworks. Thus, we will turn attention to the primary focus of this chapter to obtain an appreciation for the manner by which subnets are created and employed as well as how their utilization facilitates network management, permits a reduction in the size of routing tables, minimizes the waste of IP address space, and can result in an enhanced level of network performance. From the preceding you might obtain the impression that subnetting is equivalent to raising the flag on July 4 and eating apple pie. As we will shortly note, while subnetting might not be that patriotic, in a networking environment, its use is as common as either of the two activities mentioned for Independence Day.

The Subnetting Process

The subnetting process involves the subdivision of the two-level hierarchy of Class A, B, and C networks into a three-level hierarchy. This subdivision results in using the host portion of each 32-bit network address as a subnet number field and a reduced host field.

The top portion of Exhibit 3 illustrates a general comparison between the two-level hierarchy normally used by Class A, B, and C addresses and the three-level subnet hierarchy. The lower portion of Exhibit 3 provides specific details concerning the subdivision of the host address portion of Class A, B, and C IP addresses. Note that because the network portion of the 32-bit IPv4 address increases from 1 byte for a Class A address, to 2 bytes for a Class B address, and 3 bytes for a Class C address, the host portion of each address decreases from 3 to 2 to 1 byte. This means that although you can subnet Class A, B, or C networks, the number of bits available for the creation and the number of possible subnets you can create decreases from a Class A address to a Class B address to a Class C address.

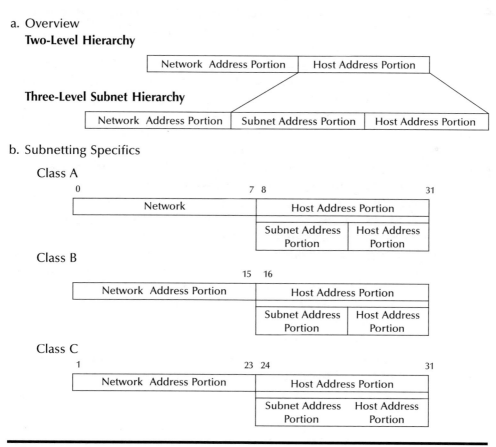

a. Overview
  Two-Level Hierarchy

Network Address Portion	Host Address Portion

  Three-Level Subnet Hierarchy

Network Address Portion	Subnet Address Portion	Host Address Portion

b. Subnetting Specifics

  Class A

  Class B

  Class C

Exhibit 3. The Three-Level Subnet Hierarchy Compared to the Two-Level Network Class Hierarchy

Through the process of subnetting, a Class A, B, or C network address can be divided into different subnet numbers, with each subnet used to identify a different network internal to an organization. Since the network portion of the address remains the same, the route from the Internet to any subnet of a given IP network address is the same. This means that routers within the organization must be able to differentiate between different subnets, but routers outside the organization consider all subnets as one network.

To illustrate the subnet process as well as obtain an appreciation for how it facilitates the use of IP addresses in a less wasteful manner and reduces routing table entries, we will examine the process. In doing so, the concept of masking and the use of the subnet mask will be discussed, both of which are essential to the extension of the network portion of an IP address beyond its network portion of the address.

Illustrative Example

To illustrate the manner by which subnetting occurs, assume our organization was issued the Class C address of 198.78.46.0. Further assume that instead of having one network, our organization reexamined the location of workstations and servers and decided that it would be optimal to install cabling for five separate networks within a building. When we called our ISP and asked for four more Class C addresses, they asked us how many stations we planned to place on each network. We told them we anticipated installing between 15 and 20 stations on each network. Persons at the ISP in charge of assigning addresses then told us that although we could apply for four additional Class C addresses, doing so would waste precious IP address space since each Class C device supports a maximum of 254 devices. In addition, when our internal network is connected to the Internet, entries for four additional network addresses would be required in a number of routers in the Internet in addition to our organization's internal routers that interconnect different organizational locations. Due to this, our ISP suggested we would be wise to use subnetting instead of requesting four additional Class C addresses. Thus, we decided to consider how to divide the host portion of the assigned Class C IP address space into a subnet number and a host number. Because we need to support five networks at one location, we must use a minimum of 3 bits from the host portion of the IP address as the subnet number. Because a Class C address uses one 8-bit byte for the host identification, this means that a maximum of 5 bit positions can be used (8 − 3) for the host number. Assuming we intend to use the 198.78.46.0 network address for our subnetting effort, we would construct an extended network prefix based on combining the network-portion of the IP address with its subnet number.

Subnet Creation Process

Exhibit 4 illustrates the creation of 5 subnets from the 198.78.46.0 network address. The top entry in Exhibit 4, which is labeled *Base Network*, represents the Class C network address with a host address byte field set to all zeros.

Exhibit 4.   Creating Extended Network Prefixes via Subnetting

Base Network:	<u>11000000.01010000.00101110</u>.00000000 = 198.78.46.0
Subnet #0:	<u>11000000.01010000.00101110.000</u>00000 = 198.78.46.0
Subnet #1:	<u>11000000.01010000.00101110.001</u>00000 = 198.78.46.32
Subnet #2:	<u>11000000.01010000.00101110.010</u>00000 = 198.78.46.64
Subnet #3:	<u>11000000.01010000.00101110.011</u>00000 = 198.78.46.96
Subnet #4:	<u>11000000.01010000.00101110.100</u>00000 = 198.78.46.128

Because we previously decided to use 3 bits from the host portion of the Class C IP address to develop an extended network prefix, the five entries in Exhibit 4 below the base network entry indicate the use of 3 bits from the host position in the address to create extended prefixes that identify five distinct subnets created from one IP Class C address. To the Internet, all five networks appear as the network address 198.78.46.0, with the router at an organization responsible for directing traffic to the appropriate subnet. It is important to note that externally to the organization, i.e., to the Internet, there is no knowledge that the dotted decimal numbers shown in the right column represent distinct subnets. This is because the Internet views the first byte of each dotted decimal number and notes that the first 2 bits are set. Doing so tells routers on the Internet that the address is a Class C address for which the first 3 bytes represent the network portion of the IP address and the fourth byte represents the host address. Thus, to the outside world, address 198.78.46.32 would not be recognized as subnet 1. Instead, a router would interpret the address as network 198.78.46.0 with host address 32. Similarly, subnet 4 would appear as network address 198.78.46.0 with host address 128. However, internally within an organization, each of the addresses listed in the right column in Exhibit 4 would be recognized as a subnet. To visualize this dual interpretation of network addresses, consider Exhibit 5 that illustrates the Internet versus the private internal network view of subnets.

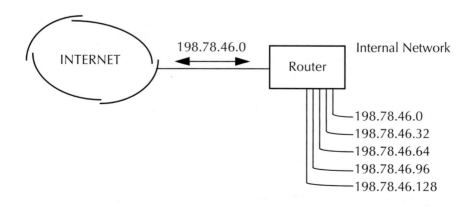

Exhibit 5.   Internet versus Internal Network View of Subnets

External View

Exhibit 5 illustrates the manner by which traffic flowing from the Internet to the internal organizational network is viewed, i.e., each IP datagram destined to the 198.78.46.0 network is first routed toward the organizational network connected to the router. Even though that router is shown as being connected to 5 subnets, to traffic on the Internet side of the router, there is no knowledge of the existence of a subnet structure. Thus, routers on the Internet only have to maintain one network address that permits simpler routing tables to be maintained.

As we might logically assume from our prior discussion of Class C addresses, any address with the network prefix 198.78.46.0 will be routed to the corporate router. However, although we noted how subnet addresses are formed, we have yet to discuss how we assign host addresses to devices connected to different subnets, or how the router can break down a subnet address so it can correctly route traffic to an appropriate subnet. We will now turn our attention to these topics.

Host Addresses on Subnets

We previously subdivided the host portion of a Class C address into a 3-bit subnet field and a 5-bit host field. When we consider the number of hosts we can assign to a subnet, we must consider two restrictions, referred to as the all-0s subnet and the all-1s subnet. When subnetting was originally defined in RFC 950, it prohibited the use of the all-0s and the all-1s subnets. The rationale for prohibiting the use of an all-0s subnet was to alleviate the situation where a classful router could become confused. Here the term classful represents any IP address that is subdivided into a network and host field. Thus, classful references Class A, B, and C IPv4 addresses.

To appreciate the manner by which a classful router can become confused, let us first digress a bit and review the term *prefix-length*, which represents another method of indicating the network portion of a classful address. The prefix-length is a number that identifies the length of the network portion of an IP address or extended network portion of an IP address. The use of a prefix-length includes a forward slash (/) that prefixes the prefix-length. For example, a conventional Class C network address would be written as follows to indicate that the network was 24 bits in length: 2.x.y.z/24, where w, x, y, and z represent the dotted decimal digits of a Class C address.

Let us return to the manner by which a classful router could become confused by the use of an all-0s subnet, a route to an all-0s subnet, and a route to the entire network. Unfortunately certain routing protocols, such as RIP-1, do not provide a mask or prefix-length with each route. When this occurs, the routing advertisements, for example, for the network 205.131.175.0/24 and its zero subnet, 205.131.175.0/27, are exactly the same, i.e., both the network and the subnet have the identical address of 205.131.175.0. Thus, unless the router knows the prefix-length or mask, it cannot tell the difference between a route to the all-0s subnet and the route to the entire network.

Although an all-1s subnet represents a broadcast address on a subnet, its use would also create confusion if used without a prefix-length in a routing table entry. For example, the broadcast address of the Class C network 205.131.175.0 is 205.131.175.255. That broadcast address is the same for the entire network (205.131.175.0/24) and the all-1s subnet (205.131.175.224/27). Thus, the host portion of a subnet field cannot be all 0s or all 1s.

Because the host field of an IP address cannot contain all 0 bits or all 1 bits, the use of 5 bits in the host portion of each subnet address means that each subnet can support a maximum of $2^5 - 2$ or 30 addresses. Thus, we could use host addresses 1 through 30 on each subnet.

Exhibit 6 illustrates the assignment of host addresses for subnet 3 whose creation was previously indicated in Exhibit 4. In examining Exhibit 6, note that we start with the subnet address 198.78.46.96 for which the first 3 bits in the fourth byte are used to indicate the subnet. Then, we use the remaining 5 bits to define the host address on each subnet. Thus, the address 198.78.46.96 represents the third subnet, while addresses 198.78.46.97 through 198.78.46.126 represent hosts 1 through 30 that can reside on subnet 3. Note that because subnets are numbered from 0, subnet 3 actually represents the fourth subnet out of the five subnets shown in Exhibit 5.

Although we now have an appreciation for creating subnets and host addresses on subnets, an unanswered question is how do devices on a private network recognize subnet addressing? For example, if a packet arrives at an organizational router with the destination address 198.78.46.97, how does the router know to route that packet onto subnet 3? The answer to this question involves what is known as the subnet mask.

The Subnet Mask

The subnet mask represents a mechanism that enables devices on a network to determine the separation of an IP address into its network, subnet, and host portions. To accomplish this, the subnet mask consists of a sequence of 1-bits that denotes the length of the network and subnet portions of the IP network address associated with a network. For example, assume our network address is 198.78.46.96 and we want to develop a subnet mask that can be

---

Exhibit 6. Assigning Host Addresses by Subnet

Subnet #2	11000000.01010000.00101110.01100000 = 198.78.46.96
Host #1:	11000000.01010000.00101110.01100001 = 198.78.46.97
Host #2:	11000000.01010000.00101110.01100010 = 198.78.46.98
Host #3:	11000000.01010000.00101110.01100011 = 198.78.46.99
!	! ! ! !
!	!
!	!
Host #30	11000000.01010000.00101110.01111110 = 198.78.46.126

used to identify the extended network. Because we previously used 3 bits from the host portion of the IP address, the subnet mask would become:

11111111.11111111.11111111.11100000

Similar to the manner by which IP addresses can be expressed using dotted-decimal notation, we can also express subnet masks using that notation. Doing so, we can express the subnet mask as:

255.255.255.224

The subnet mask tells the device examining an IP address that bits in the address should be treated as the extended network address consisting of network and subnet addresses. Then, the remaining bits that are not set in the mask indicate the host of the extended network address. However, how does a device determine the subnet of the destination address? Because the subnet mask indicates the length of the extended network to include the network and subnet fields, knowing the length of the network portion of the address provides a device with the ability to determine the number of bits in the subnet field. Once this is accomplished, the device can determine the value of those bits, which indicates the subnet. To illustrate this concept, use the IP address 198.78.46.97 and the subnet mask 255.255.255.224, with the latter used to define a 27-bit extended network. The relationship between the IP address and the subnet mask is shown below:

| IP Address: | 198.78.46.97 | 11000000.01010000.00101110.01100001 |
| Subnet Mask: | 255.255.255.244 | 11111111.11111111.11111111.11100000 |

In examining the above IP address and subnet mask, note that the first 2 bits in the IP address are set. This indicates a Class C address. Because a Class C address consists of 3 bytes used for the network address and 1 byte for the host address, this also means the subnet must be 3 bits in length (27 − 24). Thus, bits 25 through 27, which are set to 011 in the IP address, identify the subnet as subnet 3. Because the last 5 bits in the subnet mask are set to zero, this means that those bit positions in the IP address identify the host on subnet 3. Because those bits have the value 00001, this means the IP address references host 1 on subnet 3 on network 198.78.46.0.

Configuration Examples

When configuring a workstation or server to operate on a TCP/IP network, most network operating systems require you to enter a minimum of three IP addresses and an optional subnet mask or mask bit setting. The three IP addresses include the IP address assigned to the workstation or server, the IP address of the gateway or router which is responsible for relaying packets with a destination that is not on the local network to a different network, and a name resolver which, as previously noted in Chapter 4, is also referred to as the Domain Name Server or DNS. The latter is a computer that is responsible for translating near English mnemonic names assigned to computers into IP addresses.

Internet Protocol Properties

Exhibit 7 illustrates the first two configuration screens in a series of screens displayed when you select Local Area Connection Properties by right clicking on the LAN icon located in the Windows 2000 control panel. Note the third component listed in the left dialog box shown in Exhibit 7 is labeled *Internet Protocol (TCP/IP)*. If you highlight that component and click on the button labeled Properties, the result will be the display of the dialog box shown in the right portion of Exhibit 7.

That dialog box, which is labeled Internet Protocol (TCP/IP) Properties, provides you with the ability to either enter a static IP address for your workstation or server or obtain an IP address automatically. Concerning the latter, this requires your organization to operate a Dynamic Host Configuration Protocol (DHCP) server. That server will receive a request from the DHCP client, which your computer automatically becomes when the button to the left of the entry *Obtain an IP Address Automatically* is checked. When this action occurs, the server, assuming all IP addresses available have not been

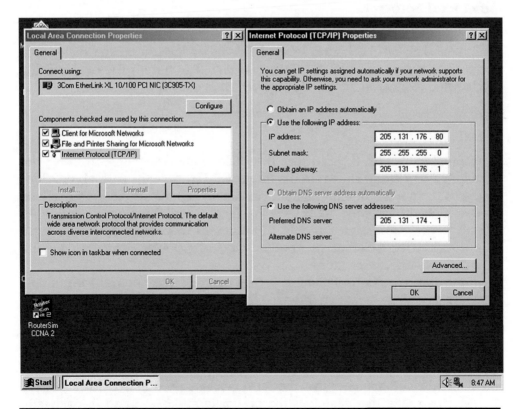

Exhibit 7. Configuring a Static IP Address for a Workstation

used, will lease an IP address to your computer. That IP address will be associated with your computer either until the lease expires or you disconnect from the network.

Because we will assume our organization uses static IP addresses, we clicked on the button to the left of the label *Use the Following IP Addresses* and entered the IP address for the workstation, the subnet mask, and default gateway IP address. Note that the subnet mask has the value 255.255.255.0. Because the IP address of 205.131.176.80 represents a Class C address, this means that the first 3 bytes externally to the organization represent the network address while the fourth byte represents the host address. Because the subnet mask only has ones set for the first 3 bytes (255.255.255.0), this means the extended network address covered by the subnet mask only covers the network address. Thus, no subnetting is in effect on the 205.131.176.0 network.

In the lower portion of Exhibit 7, note the setting of the option labeled *Use the Following DNS Server Addresses*. This setting provides you with the ability to inform or point your computer to the IP address where the local name resolution process occurs. Note that under Windows 2000 you can initially set the address for two DNS servers — a preferred server and an alternate DNS server.

Advanced Settings

In continuing our explanation of the configuration of TCP/IP addresses for a computer, click on the button labeled *Advanced* located on the dialog box in the right portion of Exhibit 7. Doing so results in the display of the dialog box labeled *Advanced TCP/IP Settings* which is shown in Exhibit 8.

IP Settings Tab

In examining Exhibit 8, note that the dialog box has four tabs, with the tab labeled *IP Settings* shown in the foreground by default. This tab provides you with the ability to manipulate a previously entered IP address and one or two default gateway addresses. In fact, using the upper portion of Exhibit 8, you can assign multiple IP addresses to your computer because it is possible for a server or another type of popular device to require multiple network connections.

In the lower portion of Exhibit 8, you can add the address for multiple gateways as well as define an interface metric for each gateway. The interface metric represents the cost of using the route associated with the connection and becomes the value in the metric column for those routes in the IP routing table. By assigning a higher interface metric to alternate gateways, you in effect inform Windows to attempt to initially use the gateway with the lowest interface metric. Remember that the term *gateway* used in Windows actually represents a router that examines the destination address in an IP datagram. If the destination address is not on the local network, the router consults its routing table to determine the port to relay the datagram toward its destination.

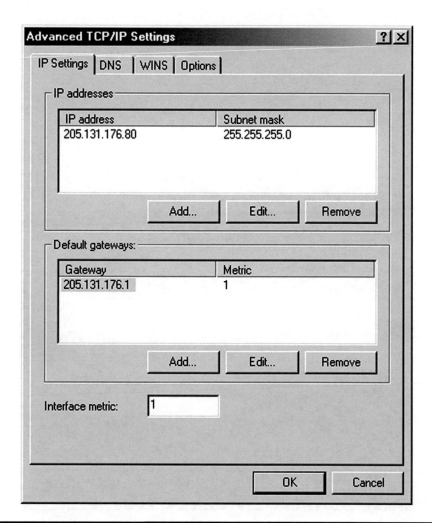

Exhibit 8. The IP Settings Tab in the Advanced TCP/IP Settings Dialog Box

DNS Tab

Similar to obtaining the ability to manipulate IP addresses through the use of the IP Settings tab, the DNS tab provides you with the ability to add, remove, and edit DNS server addresses. Exhibit 9 illustrates the DNS tab placed in the foreground of the dialog box labeled Advanced TCP/IP Settings.

In examining the upper portion of Exhibit 9, note that this portion of the tab provides you with the ability to add, edit, and remove the IP addresses of DNS servers. The lower portion of Exhibit 9 permits you to type the DNS suffix you want to use, such as .edu or .gov. Note that the default setting for the local primary DNS suffix is the same as the Windows 2000 Active Directory domain name. Thus, changing your DNS suffix will not affect your domain membership; however, it could prevent other users from finding your computer on the network.

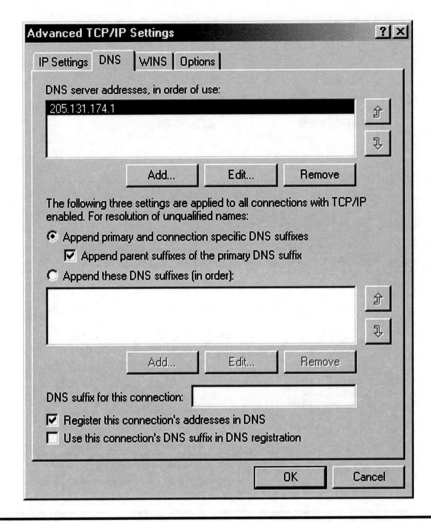

Exhibit 9. The DNS Tab in the Advanced TCP/IP Settings Dialog Box

WINS Tab

The third tab in the Advanced TCP/IP Setting dialog box is labeled WINS. This is an acronym for Windows Internet Name Service. When Windows was first developed, its networking capability relied on a protocol referred to as NetBIOS, an acronym for Network Basic Input/Output System. In general, under NetBIOS, computers were named using a single part such as Gil, Fred, or Gail. In comparison, under TCP/IP, the naming convention results in a computer name having a host name and a domain name, which together form a fully qualified domain name. Fortunately, NetBIOS computer names are compatible with DNS host names, which permits interoperability between the two. However, the first method supported by Windows to map computer names to IP addresses relied on the use of text files. For NetBIOS names, the file LMHOSTS was used, while the file HOSTS was used for DNS names. While TCP/IP now uses DNS and the HOSTS file is essentially history, some networks

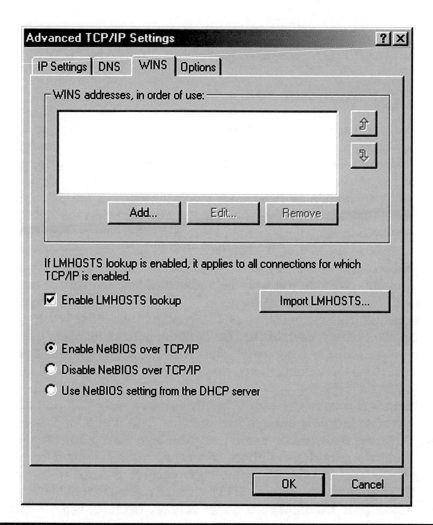

Exhibit 10. Defining the Location of One or More WINS Servers

still use WINS. Thus, the WINS tab, which is shown in Exhibit 10, provides you with the ability to retain the use of the LMHOSTS file.

In examining Exhibit 10, the top portion of the WINS tab provides you with the ability to enter the addresses of any WINS servers that Windows 2000 TCP/IP queries to resolve NetBIOS names. The WINS servers will be queried in the order they are entered in the top portion of the tab. The lower portion of the WINS tab provides you with the ability to specify the network connection. Here you can use NetBIOS over TCP/IP, which is the default setting when an IP address is manually configured. This setting is required when your computer communicates by name with computers that use earlier releases of Windows. If you only communicate with the Internet and other computers that support DNS, you can disable this option. Note that the third option at the bottom of the WINS tab provides you with the ability to have your computer obtain its network connection settings NetBT (NetBIOS over TCP/IP) and WINS from a Dynamic Host Configuration Protocol (DHCP) server. When an

IP address is obtained from a DHCP server, the computer will use NetBT (NetBIOS over TCP/IP) configuration settings as supplied by the DHCP server.

Options Tab

The last tab in the Advanced TCP/IP Settings dialog box is labeled *Options*. This tab, which is shown in Exhibit 11, lists optional TCP/IP configuration settings that are available for use by the computer you are configuring. While entries in this tab are not required for basic networking, they provide additional functionality. In examining Exhibit 11, you will note that the left portion of the Exhibit 11 contains the display of the Options tab while the right portion of the display illustrates the available settings for IP security. You can either enable or disable the use of IP security.

The second optional setting under the Options tab concerns TCP/IP filtering. This option provides you with the ability to define the type of traffic that can flow through your LAN adapters.

To conclude our examination of the configuration of a Windows 2000 computer, we will turn attention to TCP/IP filtering. If we highlight the TCP/IP filtering entry from the Options tab shown in the left portion of Exhibit 11 and then click on the button labeled *Properties*, the dialog box labeled *TCP/IP Filtering* will be displayed. An example of this dialog box into which I added a restriction is shown in Exhibit 12. In this example, only TCP port 80 traffic will be permitted to flow through the interface. TCP port 80 represents Web traffic. However, it should be noted that prior to enabling any filtering, you should carefully consider the effect, if any, on your computer operations. By restricting TCP to port 80 traffic, you preclude the ability to establish secure HTTP connections because HTTPs, which represents the Secure Socket Layer protocol for handling secure transactions, operates using a different TCP port number.

Subnet Design Considerations

Until now, the primary effort in this chapter was to obtain a basic understanding of the process involved in subnetting and examine the manner by which IP addressing information is entered into a TCP/IP configuration screen. Concerning the latter, we previously noted that under Windows we may need to carefully consider not only the entry of basic IP addressing information to include the applicable subnet mask, but in addition a potential range of additional data that can range in scope from specifying the location of one or more WINS servers to the filtering of data at the LAN interface. Because the primary focus of this chapter is on subnetting, we will return to this topic. In doing so, examine some of the key questions you should answer prior to developing a subnet design structure or architecture for your organization.

Exhibit 13 lists four key questions needing answers prior to designing a subnet structure for your organization. In examining the entries in Exhibit 13, note that the first two questions concern the current and possible expansion

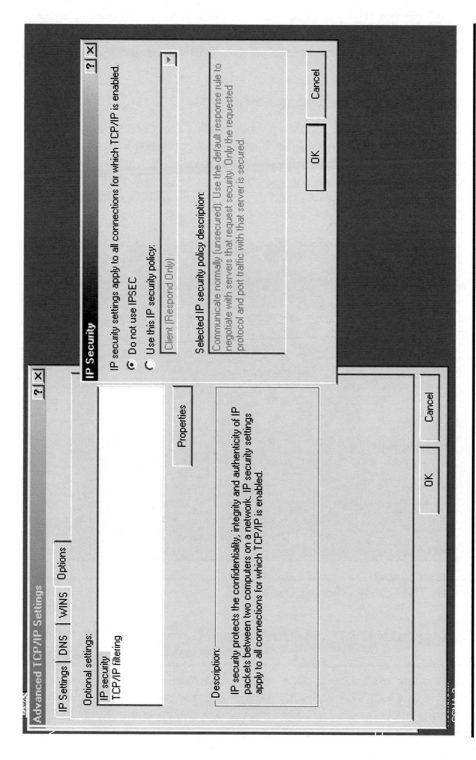

Exhibit 11. Enabling or Disabling the Use of IP Security in the Options Tab

of your organization's separate networks. For example, suppose your organization has individual LANs on each floor on five floors of a 16-story building. In addition, assume that management has options to lease two additional floors as they become available. Thus, while the answer to the first question in Exhibit 13 would be 5, the answer to the second question would be 7.

Considering the Number of Subnets

The first step in the subnet planning process uses the maximum number of subnets that will be required in the future as a preliminary design criteria. The reason the term *preliminary* is used is because if the number of subnets required in the future exceeds the capability of the current network address capability for subnetting, you will then either require one or more IP addresses or a different class of IP address.

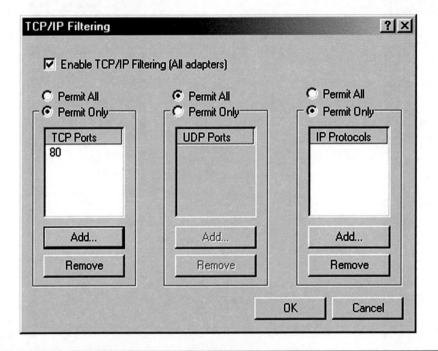

Exhibit 12. The TCP/IP Filtering Dialog Box

Exhibit 13. Questions that Need to Be Answered Prior to Performing a Subnet Design

- How many total subnets does your organization currently require?
- How many total subnets will your organization need in the future?
- What is the maximum number of hosts that will currently reside on the subnets your organization will operate?
- What is the maximum number of hosts that may reside on the largest subnet your organization may operate in the future?

Assume our organization will require 7 subnets. We would round this number up to the nearest power of two. Thus, since 2^3 is 8, this means we will require 3 bit positions to define the relevant subnets.

Host Considerations

Next, we need to consider the number of hosts that will reside on the largest subnet today and in the future. We need to consider the maximum number of hosts to determine if the reduced portion of the host address field is sufficient in length to support the largest number of hosts that will be located on a subnet. For example, if we are working with a Class B address, the host portion of that address is 2 bytes or 16 bits in length. In comparison, if we are working with a Class C address, the host portion of that address is 1 byte or 8 bits in length. Thus, if we need 7 subnets, this means we would use 3 bit positions from the host field for subnetting, leaving 13 bits in a Class B address or 5 bits in a Class C address for the reduced host portion of the original host field. If we assume that the largest subnet will have 30 hosts, we then need to determine the number of bits required to support 30 hosts. Because $2^5 - 2$ is 30, we would then require 5 bit positions in the reduced host field to support 30 hosts. Remember, the reason we subtract 2 is because we cannot use a subnet of all 0s or one of all 1s. Thus, for both a Class B and a Class C network we could easily accommodate up to 7 subnets with up to 30 hosts per subnet. If the largest number of hosts would exceed the number of available bit positions, you could either consider moving some hosts onto a different subnet or requesting another IP address. Prior to illustrating another example of subnetting, we will digress briefly and review in more detail another method used to reference address classes. This method is referred to as the prefix method.

The Prefix Method of Address Reference

Remembering our discussion of classful addressing, we noted that a Class A address has an 8-bit byte used for the network portion of the address. Similarly, a Class B address uses 16 bits for the network portion, while a Class C address uses 24 bits. As noted earlier in this chapter, another method commonly used to define the network portion of a classful IP address is via the use of a forward slash (/) character followed by a numeric that indicates the length of the network portion of the address. Thus, we can describe Class A, B, and C networks as follows:

Class A address w.x.y.z/8
Class B address w.x.y.z/16
Class C address w.x.y.z/24

where w.x.y.z. represents applicable classful dotted decimal numbers. Note that we can also apply the use of a prefix to the extended network portion of a network. For example, assume you subnetted the network 205.131.176.0 to permit up to 8 subnets through the use of a 3-bit network extension. You could then reference the 27-bit length extended network either as 127 when using the prefix notational method or as a subnet mask of 255.255.25.224.

Now that we have an appreciation for the manner by which we can express the network portion of a classful IP address, we will continue our probe into the wonderful world of subnetting with another example.

Developing a Subnet Architecture

For another example of subnetting, assume our organization was assigned the Class C IP address 205.131.175.0. We can indicate that this network address is a Class C address via the prefix method by defining it as 205.131.176.0/24.

Assume our organization requires 6 subnets, with the largest number of hosts on a subnet determined to be 28. The first step in our subnetting process is to define the number of bits that will be required to define 6 subnets. Because each bit position represents a power of two we would use 3 bits. Because 2^3 is 8, the use of 3 bits results in the ability to define 8 subnets, or 2 above our requirements. Note that if we used 2 bit positions, we could only define 2^2 or 4 subnets, 2 less than our requirements.

The use of 3 bit positions for the subnet results in a /27 extended network prefix. The subnet mask then becomes 225.225.225.224 in dotted decimal notation. Exhibit 14 illustrates the relationship between the 205.131.176.0/24 network we want to subnet and the extended network prefix or subnet mask. Note that the use of a 27-bit extended network prefix results in 5 bits being available to define host addresses on each subnet. While 2^5 results in 32 individual IP addresses being available, we cannot use an all-0s or an all-1s host address. Thus we are limited to $2^5 - 2$ or 30 assignable host addresses on each subnet.

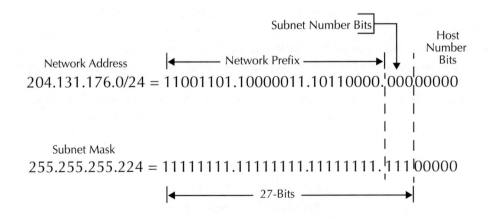

Exhibit 14.   The Relationship between the Class C Network 205.131.176.0 and Its Subnet Mask Required to Support Eight Subnets

Exhibit 15.   Subnetting the 205.131.176.0/24 Network into Eight Subnets

Base network:	<u>11001101.10000011.10110000</u>.00000000 = 205.131.176.0/24
Subnet 0:	<u>11001101.10000011.10110000.000</u>00000 = 205.131.176.0/27
Subnet 1:	<u>11001101.10000011.10110001.001</u>00000 = 205.131.176.32/27
Subnet 2:	<u>11001101.10000011.10110001.010</u>00000 = 205.131.176.64/27
Subnet 3:	<u>11001101.10000011.10110001.011</u>00000 = 205.131.176.96/27
Subnet 4:	<u>11001101.10000011.10110000.100</u>00000 = 205.131.176.128/27
Subnet 5:	<u>11001101.10000011.10110000.101</u>00000 = 205.131.176.160/27
Subnet 6:	<u>11001101.10000011.10110000.110</u>00000 = 205.131.176.192/27
Subnet 7:	<u>11001101.10000011.10110000.111</u>00000 = 205.131.176.224/27

Based on the use of 3 bit positions for the subnet number, the 8 subnets would be numbered from decimal zero (000_2) through decimal seven (111_2). Exhibit 15 illustrates the base network and the subnet numbers for the 8 subnets we can use. In examining the entries in Exhibit 15, note that all subnets other than subnet 0 are multiples of 32. As a general mechanism for checking, if your subnets are correct, you can verify your computations by ensuring they are all multiples of the subnet 1 address' last byte value. Also note that the underlined portion of each address identifies the extended network prefix.

Configuration Issues

Now that we have identified the subnet mask for each subnet, we need to consider several configuration issues to ensure a router can identify each subnet. In addition, we need to ensure that each station on each subnet recognizes their correct address. To illustrate configuration issues, assume our internal network is as shown in Exhibit 16. In this example we are using a router with seven ports, one serial and six Ethernet ports, since our initial requirement was to support 6 subnets. Note that to provide a bit of correspondence between Ethernet port numbers on the router and subnets, we connected router ports to an applicable subnet, although this is not a necessity. Thus, router port E0 is shown cabled to subnet 0, port E1 is shown cabled to subnet 1, and so on. The lower portion of Exhibit 16 indicates the subnet mask in prefix notation that would be used when you assign an IP address to each router port. That subnet mask would also be assigned to each host on the subnet connected to the applicable router port.

Defining Subnet Broadcast Addresses

In concluding our example of subnetting, we will discuss how we can easily note the broadcast address for each of the subnets in our network. For example, consider subnet 0. The broadcast address for subnet 0 is the all 1s host address or:

11001101.10000011.10110000.00011111 = 205.131.175.31

Note that the broadcast address for subnet θ is exactly one less than the base address for subnet 1. As indicated in Exhibit 15, the base address for subnet 1 is 205.131.175.32.

As we examine the broadcast address for other subnets, we will note that for subnet n, the broadcast address is always one less than the base address for subnet n + 1. For example, consider subnet 5 shown in Exhibit 15. The broadcast address for subnet 5 would be:

$$11001101.10000011.10110000.10111111 = 205.131.176.191$$

Once again, the broadcast address is one less than the base address for the next higher subnet.

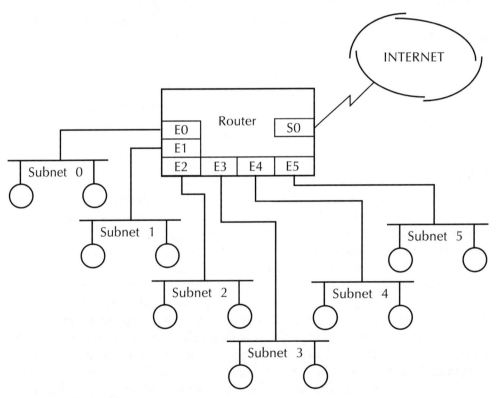

Subnet mask used by router ports:

E0   205.131.176.0/27
E1   205.131.176.32/27
E2   205.131.176.64/27
E3   205.131.176.96/27
E4   205.131.176.128/27
E5   205.131.176.160/27

Exhibit 16.   Configuring a Router to Support Six Subnets

Using Host/Subnet Quantity Tables

In concluding our coverage of subnets in this chapter, we will turn attention to the use of a series of tables that can facilitate the subnetting process. The first table we will examine, which is contained in Exhibit 17, indicates the relationship between the number of subnets, number of hosts, and the subnet mask for Class A networks. In examining the entries in this table, note that because the composition of the second byte defines the dotted decimal number used in a network extension, its composition is included in binary in the last column in the table. Also note that you can continue to adjust Class A addresses into additional subnets, each with a lesser number of hosts by using a longer network prefix. This fact will become obvious when we examine some commonality between Class A and Class B networks when we use a network extension in the third dotted decimal position of a subnet mask.

While you can subnet a Class A network into 254 separate entities through the use of an extension to the network number in the first byte, this still results in each network being capable of having 65534 hosts. As a mechanism to further subdivide a Class A network, we can use additional bits from the third byte in a Class A address for extending the subnet mask. However, because doing this is equivalent to subnetting a Class B network, we can literally accomplish two items at one time by focusing our attention on the use of the third byte in a classful IP address for subnetting. This use of the third byte for subnetting Class A and B networks is shown in Exhibit 18.

In examining the entries in Exhibit 18, note that the use of the third byte as a mechanism to extend the network portion of Class A and Class B addresses is not all inclusive. That is, we can move the network extension into the fourth byte position and continue the adjustment of Class A and Class B networks into additional subnets, each capable of containing a maximum number of hosts that diminishes as the number of possible subnets increases. However, when we extend the network portion of a classful IP address into the fourth byte, we are also capable of creating the subnet mask for a Class C network. Thus, we can again economize on the use of table space and note the effect of extending a network address into the fourth byte position

Exhibit 17. Subnet Masks for Class A Networks

Number of Subnets	Number of Class A Hosts	Bits Required	Subnet Mask	Binary Value of Second Byte
2	4,194,302	2	255.192.0.0	11000000
6	2,097,150	3	255.224.0.0	11100000
14	1,048,574	4	255.240.0.0	11110000
30	524,286	5	255.248.0.0	11111000
62	262,142	6	255.252.0.0	11111100
126	131,070	7	255.254.0.0	11111110
254	65,534	8	255.255.0.0	11111111

with respect to all three classful networks. This information is shown in Exhibit 19.

In examining the entries in Exhibit 19, a distinction is in order concerning mathematical possible subnetting and practical subnetting. From a mathematical view, it is possible to use all bits but the last bit in a 32-bit address for an extended network. However, if you do so when using a Class C address, you wind up with 126 subnets, each having 1 host which in effect represents a useless network. Similarly, using 15 bits to extend a Class B network or 23 bits to extend a Class A network also results in the largest number of subnets; however, each subnet is also restricted to one host. Because of the preceding, the maximum number of bits used to extend Class A, B, and C network addresses are 22, 14, and 6, respectively. Thus, the last line in Exhibit 19, while interesting, is not used.

Exhibit 18.   Using the Third Byte to Subnet Class A and Class B Networks

Number of Class A Subnets	Number of Class B Subnets	Number of Class A or B Hosts per Subnet	Required Bits Class A	Required Bits Class B	Subnet Mask	Binary Value of Third Byte
1,022	2	16,382	10	2	255.255.192.0	11000000
2,046	6	8,190	11	3	255.255.224.0	11100000
4,094	14	4,094	12	4	255.255.240.0	11110000
8,190	30	2,046	13	5	255.255.248.0	11111000
16,384	62	1,022	14	6	255.255.252.0	11111100
32,766	126	510	15	7	255.255.254.0	11111110
65,534	254	254	16	8	255.255.255.0	11111111

Exhibit 19.   Using the Fourth Byte to Subnet Class A, Class B, or Class C Networks

Number of Class A Subnets	Number of Class B Subnets	Number of Class C Subnets	Number of Class A, B, or C Hosts per Subnet	Required Bits Class A	Required Bits Class B	Required Bits Class C	Subnet Mask	Binary Value of Fourth Byte
262,142	1,022	2	62	18	10	2	255.255.255.192	11000000
524,286	2,046	6	30	19	11	3	255.255.255.224	11100000
1,048,574	4,094	14	14	20	12	4	255.255.255.240	11110000
2,097,150	8,190	30	6	21	13	5	255.255.255.248	11111000
4,164,302	16,382	62	2	22	14	6	255.255.255.252	11111100
8,388,606	32,766	126	1	23	15	7	255.255.255.254	11111110

Chapter 6

CIDR and Multicast Operations

Chapter 6 will turn attention to two methods that economize on the use of IP address space. The first method, referred to as Classless Inter-Domain Routing (CIDR), was developed as a mechanism to promote the more efficient allocation of IPv4 address space. It is also referred to as supernetting. The second method we will discuss, multicasting, was primarily developed as a mechanism to reduce the amount of network traffic by permitting users desiring to access a traffic flow to join a multicast group. However, because one flow replaces individual traffic flows, multicast can also be used to conserve the use of IP address space.

CIDR

Classless Inter-Domain Routing (CIDR) can be considered to represent one of the first address conservation techniques resulting from the successful growth of the Internet. If we perform a bit of research concerning the history of the Internet, we would find that until approximately 1990, use of the Internet was primarily restricted to interconnecting research laboratories and academic institutions. While commercial use of the Internet did not begin until several years later, the proliferation of connections required for linking colleges, universities, and even community colleges began to adversely affect the availability of IP addresses. By 1992, all Class A addresses were in use and Class B address space was rapidly being depleted. As more and more classful addresses were placed into use, the number of entries required in routing tables correspondingly increased, resulting in routing table problems previously noted in the discussion of subnetting in Chapter 5. As a mechanism to provide for the more efficient allocation of IP address space, several Internet working groups were established. The efforts of these working groups were oriented toward both short-range and long-term solutions to the rapid depletion of IPv4 address space. The long-term resolution resulted in the development of IPv6 (described in Chapter 8), while the short-term solution was represented by the development of CIDR.

Overview

CIDR was officially introduced to the Internet community in September 1993 as a series of Request for Comments (RFCs). RFCs 1517 through 1520 define the operation and utilization of CIDR. Because of the manner in which CIDR operates, enabling a grouping of classful IP addresses to be handled as an entity, this technique results in the capability for the creation of a network of multiple classful networks, with the term supernet or supernetting commonly used to reference the aggregation of classful networks.

Operation

Under CIDR, the concept of classful networks, such as Class A, Class B, and Class C network addresses, is replaced with the use of a network prefix. This prefix, which is denoted by a forward slash (/), followed by a numeric indicates the length of the network prefix.

Classful Addressing

Remembering our discussion of classful addressing earlier in this book, that method of addressing requires the first 3 bits of the IP address to be examined as a mechanism to determine the length of the network portion of the address, which indirectly provides the length of the host portion of the address. As a quick review, consider the IP network address 205.131.175.0. The first byte of this address has the bit composition of 11001101. Because the first 2 bits in the first byte were set, this indicates that the address represents a Class C address. This information allows a device to be programmed to note that the first 3 bytes in a Class C address represent the network portion of the IPv4 address. Because a classful address is 4 bytes in length, this means that subtracting the 3-byte network portion of the address from the 4-byte address results in 1 byte representing the network portion of the Class C address.

Classful Limitations

The previously described method associated with determining the network and host portion of a classful address is restricted to fixed byte boundaries. In addition, each router must perform several operations to determine the network and host portion of an IPv4 address. The use of CIDR replaces the concept of classful IPv4 addresses with a network prefix that is capable of spanning byte boundaries. For example, a /18 network prefix crosses the 2-byte or 16-bit boundary of a conventional classful IPv4 address. Thus, the use of CIDR provides a mechanism to deploy arbitrarily sized networks instead of the standard length of 8-, 16-, and 24-bit network numbers associated with classful Class A, B, and C networks.

CIDR Prefix-Length Routing

Under CIDR, routing information is defined through the use of a prefix length. The use of a prefix length makes it possible for routers to employ a longest-match forwarding algorithm. This algorithm permits a single routing table entry to represent the address space of hundreds to thousands of classful routes, providing another mechanism to reduce the size of router routing tables. Since a reduction in the size of a router's routing table also facilitates the table search process, a secondary benefit of the use of CIDR is an expedited routing capability.

Utilization

CIDR transcends classful networks and is applicable for use based on address space previously allocated to Class A, Class B, and Class C networks. To illustrated this concept, consider the use of a /21 prefix. In doing so, assume we are using the /21 prefix with a Class A network address 12.13.14.0 to obtain the block 12.13.14.0/21, while the /21 prefix with Class B and C addresses results in blocks 160.150.140.0/21 and 205.131.176.0/21. Through the use of a /21 prefix, we can allocate 1024 (1 K) individual addresses from each of the previously mentioned addresses. This concept is illustrated in Exhibit 1.

Address Blocks

In examining Exhibit 1 note that the CIDR prefix length represents a contiguous address space referred to as an address block. The actual number of individual addresses in a block is computed by raising 2 to the power of 32 less the value of the prefix length. For example, a prefix length of /21 results in $2^{(32-20)}$ or 4096 individual addresses falling into a CIDR block.

Because CIDR crosses classful network boundaries, it became possible to deploy address blocks that use portions of IPv4 address space that are not in use. In fact, some organizations either voluntarily or were requested to return

Exhibit 1. The CIDR Prefix Length Spans Traditional Class A, Class B, and Class C Network Numbers

Class A	12.13.14.0/21	00001100.00001101.00001110.00000000
Class B	160.150.140.0/21	10100000.10010110.10001100.00000000
Class C	205.131.176.0/21	11001101.10000011.10110000.00000000

blocks of previously assigned but unused classful addresses which enabled their use for CIDR operations.

Exhibit 2 indicates information concerning some of the most commonly deployed CIDR blocks. In examining the entries in Exhibit 2, note that CIDR can either be specified via the use of a prefix length or a dotted decimal mask. Also note that while CIDR address blocks are primarily created from Class C address space, such blocks can also be created from Class A and Class B address space. Because most Class B networks and all Class A networks were allocated prior to the deployment of CIDR, this fact explains why the majority of CIDR blocks use Class C address space.

Address Allocation

Understanding the major benefits of CIDR requires a brief discussion of the allocation of IP address space, so we will turn our attention to this topic.

IPv4 addresses in the form of blocks of address space are assigned to large Internet Service Providers (ISPs). The ISPs in turn reallocate portions of IPv4 address space to their customers. For example, assume an ISP was assigned the address block 192.12.16.0/17. This block represents 32,768 (2^{15}) IP addresses. If an ISP client had a requirement for 1000 host addresses, the allocation of a Class B network address would waste 65,536 − 1000 or 64,536 host addresses. Remember, a Class B network address uses 2 bytes for the network address and 2 bytes for the host address. Thus, 16 bits or 2 bytes permits 65,536 distinct host addresses.

Instead of allocating one Class B address, assume the ISP has considered providing your organization with four Class C addresses. While each Class C

Exhibit 2. Commonly Deployed CIDR Address Blocks

CIDR Prefix Length	Dotted-Decimal Mask	Number of Individual Addresses	Equivalent Number of Classful Networks
/13	255.248.0.0	524288 (512K)	8 Bs or 2048 Cs
/14	255.252.0.0	262144 (256K)	4 Bs or 1024 Cs
/15	255.254.0.0	131072 (128K)	2 Bs or 512 Cs
/16	255.255.0.0	65536 (64K)	1 Bs or 256 Cs
/17	255.255.128.0	32768 (32K)	128 Cs
/18	255.255.192.0	16384 (16K)	64 Cs
/19	255.255.224.0	8192 (8K)	32 Cs
/20	255.255.240.0	4096 (4K)	16 Cs
/21	255.255.248.0	2048 (2K)	8 Cs
/22	255.255.252.0	1024 (1K)	4 Cs
/23	255.255.254.0	512	2 Cs
/24	255.255.255.0	256	1 Cs
/25	255.255.255.125	128	1/2 C
/26	255.255.255.192	64	1/4 C
/27	255.255.255.224	32	1/8C

address can be used to support 254 hosts and thus reduces the amount of wasted address space, their use requires four additional routes to be placed into router routing tables. Thus, the old adage (you're damned if you do and you're damned if you don't) appears to be true concerning the use of one Class B versus four Class C addresses for satisfying your organization's IP address requirements. While the adage was true under classful IP addressing, let us examine the assignment of a CIDR address block from the ISP to our organization.

Instead of the allocation of one Class B or four Class C addresses, the ISP can allocate a block of addresses by supporting CIDR. From Exhibit 2 we note that a prefix length of /22 supports 1024 host addresses. Thus, the ISP could assign our organization a block of four contiguous Class C addresses (/24 prefix) as illustrated on the second line of Exhibit 3. Note that once the CIDR block is provided to our organization, any destination address of 192.12.20.0/22 is directed toward our organization's router. That router is configured for each of its internal interfaces with an IP address and prefix length corresponding to the right column in Exhibit 3.

Route Aggregation

In addition to providing a mechanism to more efficiently allocate blocks of IPv4 address space, CIDR conserves routing table entries. This is accomplished by the CIDR routing scheme permitting a single high-level route entry to represent many lower-level routes in router routing tables. You can view the route aggregation scheme as one similar to the telephone network, where an area code and three-digit exchange prefix can be considered to form a hierarchical structure. That is, under the area code resides numerous exchange prefixes, and under each exchange prefix resides individual phone numbers or more accurately four-digit extensions on the exchange. Thus, at a high level within the telephone network, a switch examines the dialed number and can initiate routing based on the area code. As a call is being set up, a switch serving the area code in question would look further into the dialed number and examine the prefix to route the call further along toward its destination. Similarly, the switch that serves the prefix within the area code would then examine the last four digits of the dialed number to route the call to its intended destination.

Exhibit 3.   Allocating a CIDR Address Block

ISP address block	11000000.00001100.00010000.00000000	192.12.16.0/17
Organization assigned block	11000000.00001100.00010100.00000000	192.12.20.0/21
Class C SN0	11000000.00001100.00010100.00000000	192.12.20.0/24
Class C SN1	11000000.00001100.00010101.00000000	192.12.21.0/24
Class C SN2	11000000.00001100.00010110.00000000	192.12.22.0/24
Class C SN3	11000000.00001100.00010111.00000000	192.12.23.0/24

Perhaps taking a leaf from the hierarchy of the telephone network, CIDR provides a similar routing capability. To illustrate the manner by which CIDR reduces the number of entries in routing tables, use the previously noted ISP address block of 192.12.16.0/17 shown in Exhibit 3, from which our organization was assigned the address block of 192.12.20.0/21. Further assume that the ISP our organization uses allocated a series of blocks from the 192.12.16.0/17-address block to our organization and other organizations. In doing so, let us make a further assumption that the ISP subdivided the 192.12.20.0/20-address block to satisfy two additional organizations. Exhibit 4 illustrates one possible method whereby our organization's ISP could have subdivided the CIDR 192.12.20.0/20-address block.

In examining Exhibit 4 we will assume that a portion of the ISP's address block 192.12.16.0/20 was allocated to our organization which is labeled organization A. This aggregation enables our organization, organization A, to aggregate its four 124 networks via 192.12.16.0/21. If you examine Exhibit 4, you will note that the first 21-bit positions, which represent the prefix length used for routing to organization A in Exhibit 4, define the path where the longest match includes coverage of the first 21 bits. If we focus our attention on the third byte, we would note that the values for decimal 20 through 23 have the common composition 00010 in the first 4-bit positions of the third byte which permits aggregation via 192.1.16.0/21.

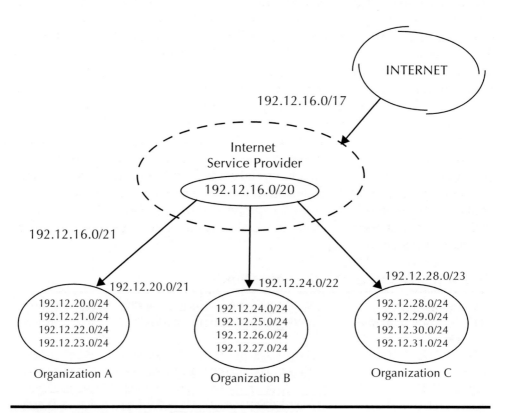

Exhibit 4. An Example of the Use of CIDR

Similarly, for organization B, whose network values range from 192.12.24.0/24 through 192.12.27.0/24, the common bit composition of the first five positions in the third byte becomes 00011, resulting in the need to match 6 bits in the third byte to differentiate one network from another. Thus 16 + 6 results in 192.12.24.0/22 becoming the common route advertisement.

Continuing our examination of Exhibit 4, note that organization C has four separate blocks of addresses that are aggregated into a single routing address. To provide a path to organization C requires matching through one additional bit position. Thus, the prefix length becomes 123 for the aggregation while the network address prefix becomes 192.12.28/23.

Based on the preceding, the CIDR address 192.12.16.0/20 is the only address that needs to be advertised from the ISP to the Internet. Once a datagram with a destination address within the 192.12.16.0/20 block arrives at the ISP, the ISP must determine where the datagram must be routed. To accomplish this task, ISP routers compare the destination address by extending a bit-matching process until a longest prefix-length match occurs. Thus, if a match occurs using a prefix length of /21, the datagram is routed toward organization A. In comparison, if a match occurs using a prefix length of /22, the datagram is routed toward organization B, while a match occurring using a prefix length of /23 results in the datagram being routed toward organization C.

Multicasting

In Chapter 2 when we discussed IPv4 addressing, we noted that a Class D address was employed for a function referred to as multicasting. We also noted that multicasting and the use of a multicast address provides a mechanism to limit the number of transmissions flowing onto a network; however, until now a detailed description of how multicasting occurs was deferred. We will now turn attention to multicasting and the use of Class D addresses in this section.

Overview

To obtain an appreciation for the role of multicasting, we will review its operation in context with the other two types of addresses supported under IPv4 — unicast and broadcast.

Unicast Addressing

The most common form of IPv4 addressing is unicast addressing. Under unicast addressing, a host communicates on an individual basis with other hosts. That is, a host requiring communications with another Class A, Class B, or Class C address adds an IP header with the applicable destination address to a TCP segment or UDP datagram to form an IP datagram. Then, the IP datagram is routed to a single destination based on the value of the destination address in the header. The top portion of Exhibit 5 illustrates unicast transmission.

a. Unicast

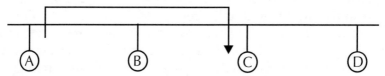

Under unicast transmission, a datagram is transmitted to a specific host.

b. Broadcast

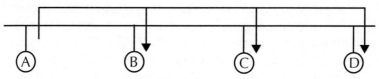

Under broadcast transmission, a datagram is transmitted to all hosts on a specified network or subnet.

c. Multicast

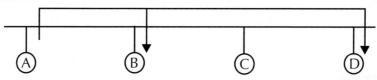

Under multicasting, a datagram is received by all members of a multicast group.

Exhibit 5.   Comparing Unicast, Broadcast, and Multicast Operations

Broadcasting

You may consider broadcasting as representing the inverse of unicast. That is, broadcasting represents the transmission of a datagram onto a network or subnet where its contents are read by each device on the network.

As we previously noted in Chapters 3 and 6, both a network and a subnet have broadcast addresses. For either situation, a broadcast address represents setting the host portion of the address to all-1s. The middle portion of Exhibit 5 illustrates broadcasting on a network.

Multicasting

Both unicast and broadcast addressing represent two extremes, i.e., under unicast addressing, datagrams are directed to a single address, while under broadcast addressing, each datagram is read by all devices on the applicable network or subnet. While having only two types of addressing is not a hardship for most types of communications operations, consider the potential effect on a network if each user pointed their browser to Victoria's Secret or a similar site to become participants in a video fashion show. If there were 200 users on your organization's network, 200 separate video flows would be directed to your organization's network. This activity would obviously be detrimental to the health of your organization's network and result in the need for the development of a more efficient method for the delivery of data to multiple recipients. That more efficient method is multicasting.

Under multicasting, a group of devices that require the ability to receive a common data flow subscribes to a multicast group. The data flow uses a Class D multicast destination address, resulting in a single flow being directed to a network. Then, each member of the multicast group reads each datagram with the applicable Class D multicast address, reducing both the amount of network traffic and the amount of bandwidth required. The lower portion of Exhibit 5 illustrates an example of a multicasting operation.

As indicated in the lower portion of Exhibit 8, multicasting can be considered to represent a *one-too-many* method of communications. Although Web casts, such as the annual Victoria's Secret fashion show, are a relatively recent phenomenon, the evolution of multicasting dates to 1992 and the initial establishment of the multicast backbone, referred to as Mbone. Through the use of a series of protocols, it becomes possible for persons to add and drop themselves from multicast groups, the latter representing a group of users that wish to subscribe to a common data flow and which will be described in additional detail later in this chapter. Other protocols permit routers to communicate with one another to enable the routing of a multicast data flow to networks with subscribers for the data flow. Once we discuss multicast groups, we will return to a discussion of the Mbone.

Multicast Groups

The ability to direct a single flow of traffic onto a network where it is read by multiple hosts resides in the creation of a multicast group. Any host in any location in IP addressing space has the ability to subscribe to a multicast group. In fact, a host can subscribe to multiple multicast groups and the receiving host can reside on the same or even a different network as the sending host. Each host in the multicast group in effect are *tuned* to the same Class D address. Thus, each host in a multicast group regardless of the location where they reside will receive the same series of IP datagrams. However, because members of a common multicast group can reside anywhere in IPv4 address space, this means that it is possible during the datagram routing process for packets to arrive out of order. Thus, members of a multicast group may need to reassemble datagrams received into their applicable order.

Use of Class D Addresses

As mentioned earlier in this section, multicasting involves the use of Class D addresses. As indicated in Chapter 3, a Class D address can have any value from decimal 224 to decimal 239 in its first byte. However, similar to other classful addresses, there are certain Class D addresses that are reserved. Exhibit 6 indicates currently reserved Class D addresses.

In addition to the reserved addresses listed in Exhibit 6, there are some multicast group addresses assigned as well-known addresses by the Internet Assigned Numbers Authority (IANA). Such multicast addresses are referred to as permanent host groups. Exhibit 7 lists three examples of multicast addresses for permanent host groups.

You can use multicasting both within an organization as well as external to an organization. When you use multicasting within a private organization, the Internet Assigned Numbers Authority (IANA) recommends the use of a Class D address in the 239.0.0.0 to 239.255.255.255 range. However, if you need to communicate a multicast over the Internet, you must request an applicable Class D address from the IANA.

The Multicast Backbone

As noted earlier in this section, the multicast backbone (Mbone) dates to 1992. However, the concept of multicasting actually dates to the 1960s when IPv4 address space was allocated, resulting in the Class D address range from 224.0.0.0 to 239.255.255.255 being served.

Exhibit 6.   Class D Reserved Addresses

Address	Utilization
224.0.0.1	Multicasting to all hosts on a subnet
224.0.0.2	Multicasting to all routers on a subnet
224.0.0.4	Multicasting to all Distance Vector Multicast Routing Protocol (DVMRP) routers
224.0.0.5	Multicasting to all Multicast Open Shortest Path First (MOSPF) routers
224.0.0.9	Multicasting to all Routing Information Protocol version 2 (RIP2) routers
224.0.0.10	Multicasting to all Interior Gateway Routing Protocol (IGRP) routers
224.0.12.0 to 224.0.12.63	Reserved for MSNBC utilization
224.0.18.0 to 224.0.18.255	Reserved for Dow Jones News Service

Exhibit 7.   Multicast Addresses for Permanent Host Groups

Permanent Host Group	Multicast Address
Network Time Protocol	224.0.1.1
RIP-2	224.0.0.9
Silicon Graphics' Dogfight Application	224.0.1.2

Structure

Mbone can be considered to represent a logical network consisting of Class D address space that is overlayed on the Internet or private network address space. Once a multicast session is assigned a Class D address, any host that wants to receive the session flow configures the multicast address associated with the session flow onto an interface designated to receive the session flow. For most devices, such as PCs that only have one interface, this action results in the assignment of a Class D address to their primary interface. Note that because each host that wishes to receive the session flow is configured with the same Class D address, the multicast can be transmitted to one address, but becomes capable of being received by multiple clients.

Router Operations

While the configuration of a Class D address on an interface is an easy to understand process, what may not be so straightforward is the manner by which routers must know where to forward datagrams containing a Class D destination address. For example, consider Exhibit 8, which illustrates a small portion of the Internet showing several routers that provide access to four networks. Let us assume two networks have one or more clients that belong to a multicast group associated with a session flow, while the other two

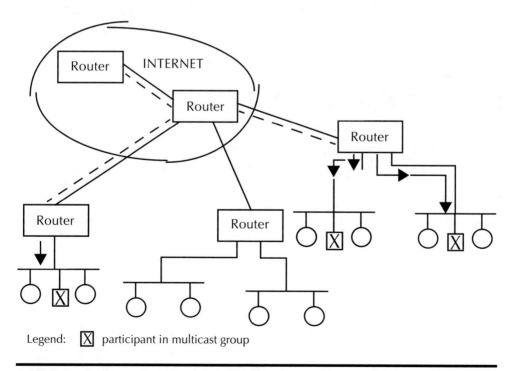

Legend: ☒ participant in multicast group

Exhibit 8. Routers Require a Mechanism to Forward Multicast Datagrams toward Networks

networks have no members of the multicast group. Because two networks shown in Exhibit 8 do not contain members of the multicast group, it would be highly inefficient to transmit session flows onto those networks. Thus, a mechanism is required to enable routers to become multicast aware and note which networks that are to receive a particular multicast session flow. That mechanism is provided by the use of an alphabet soup of protocols, referred to by the mnemonics IGMP, DVMRP, and PIM, which we will now explore.

IGMP

When a host desires to join a multicast group, it uses the Internet Group Management Protocol (IGMP) to transmit a message to the nearest multicast compliant router. IGMP messages are transported in IP datagrams as illustrated in Exhibit 9. While a 4-bit field provides the ability to define 16 types of messages, only 2 are currently defined: (1) a report and (2) a query. Hosts transmit reports to join or maintain membership in a multicast group, using the Class D address of the group. The network router receiving the report updates its multicast forwarding tables by placing the Class D address and interface from which the report was received into memory. In comparison, a multicast compliant router will periodically transmit IGMP queries to locate active multicast group members. In doing so, the IGMP host-query messages are transmitted to the *all-stations* group address 224.0.0.1. Hosts then respond to this query via IGMP report messages listing the multicast groups they would like to join.

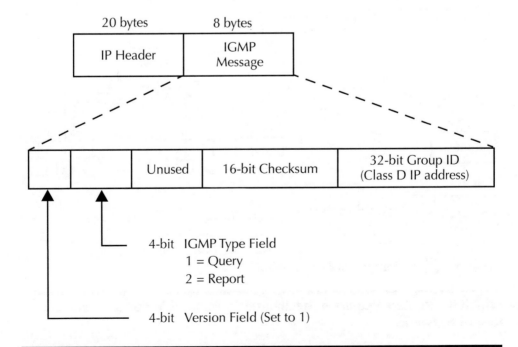

Exhibit 9.   An IP Datagram Transporting an IGMP Message

Since each multicast compliant router learns the interfaces associated with multicast groups, the distribution process for delivering Class D flows is relatively simple. That is, when a router receives a multicast datagram, it transmits a copy onto each interface listed in the forwarding table that has a group identification that matches the datagram's multicast address. Note that only one copy of each datagram is transmitted over an interface regardless of the number of members of the multicast group supported on the interface.

As previously noted, on a periodic basis, multicast compliant routers will transmit IGMP queries using the multicast address 224.0.0.1. This address represents the *all stations* multicast address. Queries have their group address set to zero. This setting causes hosts to respond with an IGMP report for each group to which they belong, providing routers with the ability to update their multicast routing tables. If there are no reports concerning a particular group on an interface, the router will delete the corresponding entry from its table and stop forwarding frames for that group over the interface. Thus, multicast frames are only transmitted via an interface when an attached network includes at least one active member of a multicast group or provides a path to a distant network that includes one or more active members of a multicast group.

Routing Protocols

Mbone routers communicate with each other using the Distance Vector Multicast Routing Protocol (DVMRP), the Protocol Independent Multicast Dense Mode (PIM-DM), or the Multicast Open Shortest Path First (MOSPF) protocols. Because the use of a particular protocol is related to the use of a spanning tree model, we will turn attention to that topic prior to discussing multicast routing protocols.

Spanning Tree Models

If we are familiar with bridging in an Ethernet environment, we know that the spanning tree is used as a mechanism to ensure that no closed loops occur in the network topology. Multicast also employs a spanning tree algorithm or, more accurately, two network structures for use by algorithms, referred to as dense mode and sparse mode, to deliver a Class D stream to its intended destinations.

Dense Mode

The use of a dense mode structure assumes that multicast group members are densely clustered in close proximity to one another in certain areas throughout a network and that bandwidth is sufficient to support the delivery of broadcasts to cluster areas. Dense mode is commonly used for large distributions, such as a Web cast of Victoria's Secret annual fashion show, where one location provides the source for a flow that will be delivered to group members located on different networks as well as a multiple number of group members located on the same network or subnet.

Sparse Mode

In comparison to dense mode, sparse mode assumes that members of a multicast group are located in small concentrated areas, where senders and receivers are separated by wide area network connections. Thus, the multicast does not have to be initially distributed to every corner of the Internet or a private intranet. Instead, sparse-mode routing is initiated by hosts transmitting requests to the source or to the next higher level multicast compliant router.

Exhibit 10 compares and contrasts dense and sparse mode operations. Under dense mode, queries flow down the tree structure to every network. In comparison, under sparse mode only requests are used to construct the applicable spanning tree.

Dense Mode Protocols

Routing protocols to include Distance Vector Multicast Routing Protocol (DVMRP), Protocol Independent Multicast Dense Mode (PIM-DM), and Multicast Open Shortest Path First (MOSPF) are best suited for a dense mode multicasting network structure.

DVMRP

The Distance Vector Multicast Routing Protocol (DVMRP) which is defined in RFC 1075 represents one of several standards available for routing IP multicast traffic. DVMRP is best suited for a dense mode topology.

When DVMRP is used, a router assumes that a host on each subnet needs to receive a Class D broadcast. This results in DVMRP employing a flooding method to transmit multicast data to its destination. This flooding method is referred to as reverse path forwarding. If a router is connected to a group of networks that do not wish to receive a particular multicast group, that router can transmit a "prune" message up the distribution tree to stop subsequent multicast datagrams from traveling to a location where no members reside.

DVMRP routers use their normal routing tables to determine if they have the best path to the next multicast compliant router. If a router determines through the use of IGMP that it has no downstream members or lacks the best path to the next router, it will request deletion from the multicast transmission. This action results in routers pruning branches where no members exist.

When DVMRP is used, broadcasts are employed to update all routers on a network. This action can result in the generation of a large amount of traffic when there are many multicast flows and networks with members of multicast groups are spread apart. Due to this, the use of DVMRP is better suited for

a. Dense Mode

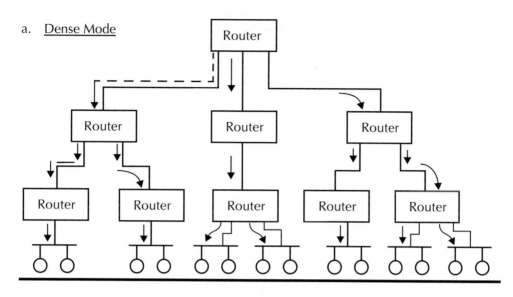

b. Sparse Mode

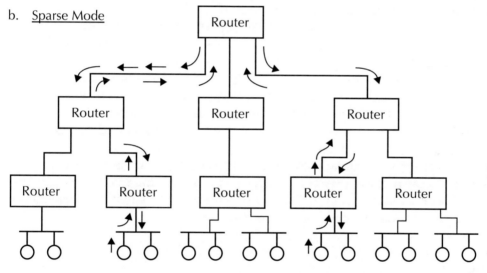

Exhibit 10. Dense Mode versus Sparse Mode Routing

supporting transmission to concentrated group members that fall into a dense mode topology. A second limitation of DVMRP is the fact that it causes a spanning tree to be created for every multicast group ID. This can result in a significant use of router resources and represents another reason why the DVMRP protocol is better suited for a dense distribution of multicast clients, which permits a reduced tree structure.

PIM-DM

The Protocol Independent Multicast Dense Mode (PIM-DM) routing protocol represents a dense version of PIM that was created by the Internet Engineering

Task Force (IETF). The goal of the development of PIM-DM was to obtain a scaleable multicast routing protocol that could operate efficiently across the multiple backbones of the public Internet as well as private networks. In addition, the PIM-DM protocol was also developed to provide independence from a particular unicast routing algorithm because it works with all existing unicast routing protocols.

Dense Mode PIM is similar to DVMRP in that it also uses a reverse path forwarding method. When a router receives a multicast datagram, it transmits the datagram onto all interfaces other than the one on which it was received.

Although PIM-DM was designed as a relatively simple protocol, its simplicity also represents a disadvantage with respect to its operation. For example, consider the manner by which this protocol operates when a datagram arrives at a multicast compliant router supporting PIM-DM. PIM-DM determines if it is using the shortest path back to the source. If it is, the router forwards packets onto all downstream interfaces until they reach networks with members of the applicable multicast group. This means it becomes possible for multiple copies of a multicast datagram to flow onto downstream networks. An example of this datagram duplication is shown in Exhibit 11. In this example, a host on the network at the top of the tree transmits a multicast datagram to Group 1 that flows to router A. Router A duplicates each datagram, transmitting packets out on interface 1 and 2 to routers B and C. Those routers duplicate the datagrams, transmitting them to routers D, E, and F.

To prevent flooding or saturating a network, PIM-DM routers need to keep track of multicast traffic and the interface on which such traffic arrives. Because router E has a host that is a member of Group 1 and received datagrams on two interfaces, it would send a prune message to either router B or C. The actual decision concerning which router should be pruned is reached through a negotiation process between routers B and C, adding additional traffic to the network. Because router F does not have any hosts that are members of Group 1, that router would send a prune message to router C.

MOSPF

A third routing protocol employed in dense-mode networks is the Multicast Open Shortest Path First (MOSPF). You can view MOSPF as an extension to the OSPF protocol that was developed to support multicast transmission. Under MOSPF, datagrams are routed over the least cost connection that has available bandwidth, with the shortest hop count to the destination used as a metric to determine the most appropriate path. This routing method permits heavily congested connections to be avoided since a highest cost in the form of hop counts can be assigned to such routes.

A router supporting MOSPF uses link-state information flooded between routers to develop a view of the entire network. Because IGMP data concerning requests and reports as well as link state information is passed between routers, the use of MOSPF will reduce available bandwidth as the number of members in a group located in a network increases. Another limitation of MOSPF concerns its reliance on OSPF domains for the routing of traffic. Because OSPF

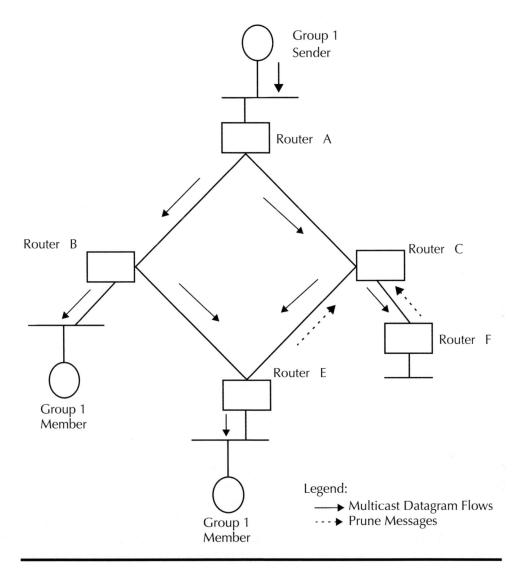

Exhibit 11.   PIM-DM Pruning Reduces Duplicate Traffic

limits transmission to group members residing on a network owned by one organization, MOSPF does not scale to the multiorganization backbone structure of the Internet. However, this protocol is still very suitable for use within an organization's intranet.

Now that we have an appreciation for the dense mode protocols, we will turn attention to the second type of multicast topology, which is referred to as sparse mode.

Sparse Mode Protocols

Unlike Dense Mode, which assumes there are group members potentially located on every network and subnet, Sparse Mode assumes that group members may be located in clusters on selective packets within a large network structure. This means it may not be necessary to transmit a multicast flow onto every network branch. Instead, Sparse Mode protocols construct routing trees by requiring other routers to participate in creating the tree. In doing so, routers operating a Sparse Mode protocol only ask to join the tree when a client in the downstream direction requests admission to a multicast group.

There are two routing protocols that support the creation of a sparse mode spanning tree topology. Those protocols are Core-Based Tree (CBT) and Protocol Independent Multicast Sparse Mode (PIM-SM). Each of these protocols constructs routing tables by requiring routers to participate in the creation of a Sparse Mode tree. That tree is created based on Sparse Mode routers receiving an admission request from a downstream host. When this action occurs, the router then informs its upstream neighbor that it wants to join a particular multicast session.

Now that we have a general appreciation for the manner by which a sparse mode topology is created, we will turn attention to the two Sparse Mode routing protocols.

CBT

Under the Core-Based Tree (CBT) routing protocol, a common tree is created that is used by each multicast group. Under CBT, a core router is used to control all data flows regardless of the source of the multicast session. Thus, the core router becomes the head of a tree structure used by all multicast groups. While the use of a common tree reduces link-state information that must be transmitted between routers, the core router can become overloaded as additional hosts on different networks join one or more multicast groups.

When CBT is used, a group member will transmit an IGMP datagram upstream to the first multicast compliant router. If that router is already receiving the multicast session, it will now transmit the flow out of the interface toward the group member. Otherwise, the router will send a request upstream toward the core router. If this upstream request reaches a router that is receiving the session, that router will add the member's router to the distribution and direct the session in the downstream direction toward the member's route. If no upstream router is currently receiving the session, the request will eventually reach the core router that will enable the session to flow in the applicable downstream direction.

Now that we have an appreciation for the manner by which CBT operates, our attention will turn to the second Sparse Mode routing protocol.

PIM-SM

The Protocol Independent Multicast Sparse Mode (PIM-SM) routing protocol replaces the core router used by the CBT protocol with rendezvous point (RP) routers. The latter represents locations where downstream routers are clustered and enables PIM-SM to create a shared tree based on the shortest path between routers. While this results in a topology similar to CBT, it provides more flexibility because each multicast group can have a different tree structure.

Under PIM-SM, it is assumed that no host wants to receive multicast traffic unless they specifically request to receive such traffic. The RPs collect information about multicast senders and make this information available to potential receivers. Thus, when a sender wants to transmit a multicast flow, it does not broadcast such data. Instead, it first transmits the data to RPs. Since receivers register with RPs, the RPs know what links the flow should be transmitted onto and the links where the flow should not flow. Thus, the use of RPs and the fact that receivers register their requirements with RPs result in any unnecessary paths being removed from the network topology.

Initially PIM-SM commences operation using a shared-tree structure. As routers join the structure, they have the ability to create their connection via a shorter path. To do so, a router transmits a message to the RP to join the shared tree. This results in the creation of a new path while the old path is disconnected. Because this action serves to reduce the size of the link-state table, it reduces the time required for routers to exchange information, minimizing latency delays. In addition, PIM-SM supports multiple RMs, permitting connections to large groups to be distributed across the network.

Exhibit 12 illustrates an example of the manner by which PIM-SM operates. In examining Exhibit 12, take notice of the term *leaf router*, which represents

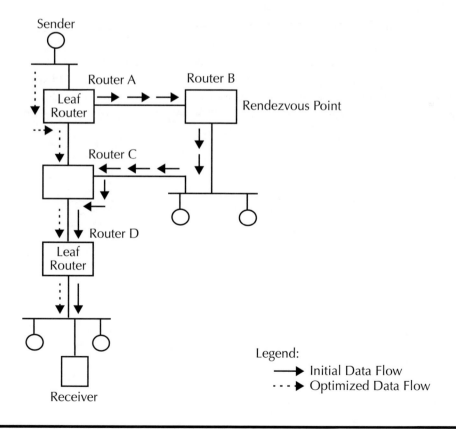

Exhibit 12.   PIM-SM Operation

routers that are directly connected either to a sender or receiver of multicast traffic. A leaf router designates one or more routers as RPs. In Exhibit 12 assume that router B was designated as a rendezvous point. Under PIM-SM, the leaf router that is directly connected to a multicast sender transmits PIM register messages on behalf of the sender to the RP. Thus, router A in Exhibit 12 transmits PIM register messages to router B. Similarly, leaf routers directly connected to receivers transmit PIM join prune messages to RPs to inform the rendezvous point about group membership. Thus, returning to Exhibit 12, router D would note it has a receiver and transmits a PIM join message to router B. As a result of this, the initial data flow in which multicast transmission flows from router A to router B to C and D would be optimized to avoid flowing to B.

Summary

Although there are numerous details to note when preparing for multicast routing, in actuality the configuration of a router in most cases is a relatively simple process. Once you determine the applicable protocol to support, you would enable the protocol for operation on an applicable interface. While the exact router command or series of commands will vary based on the router used, they typically involve a core set of commands. For example, to enable multicast routing, you would use the command *ip multicast-routing*. Then, you would add one or more router commands based on the protocol selected to support multicast routing. Thus, while many router manuals may not provide a detailed explanation of topology modes, they should provide guidance concerning the applicable commands to use to support a particular topology.

Chapter 7

Network Address Translation and IP Naming Services

Chapter 7 will focus on two topics that extend the use of IPv4 address space — (1) Network Address Translation (NAT) and (2) IP naming services. In the first section, we will examine how NAT provides us with the ability to use the RFC 1918 addresses behind a translating device, such as a firewall or router. We can then translate addresses behind the firewall or router into valid IPv4 addresses for use on the Internet. Because the translation process permits a significant reduction in the use of IPv4 public address space, NAT conserves the use of IPv4 addresses.

In the second section, we will examine the operation and utilization of IP naming services. Similar to NAT, IP naming services provide a mechanism for conserving the use of IPv4 addresses. However, instead of translating a large number of non-Internet ready addresses into a lesser number of Internet ready addresses, the use of a naming service distributes addresses on a dynamic and normally temporary basis.

Network Address Translation

During the first two decades of the existence of the Internet, its use as a mechanism to interconnect research laboratories and universities did not severely tax the use of available IPv4 address space. However, commercialization of the Internet and the growth in the World Wide Web, the literal explosion in the use of e-mail, and other applications considerably increased the rate of depletion of IPv4 address space. Although the Internet Engineering Task Force (IETF) recognized the pending demise of available IPv4 address space and implemented work on IPv6, its availability was viewed as being in the distant future. Thus, a short-term solution for extending IPv4 address space was required to enable growth of the Internet to continue. One of several short-term solutions occurred in 1994, with the publication of RFC 1631, "The IP Network Address Translator." When employed in conjunction with RFC 1918, "Address Allocation for Private Internets," it becomes possible for organizations to use the same IP addresses behind their router or firewall and map those addresses into one or more valid IPv4 addresses based on one of several methods. For example, as noted earlier in this book, RFC 1918 defines three blocks of private Internet addresses. As a review, those address blocks are:

> 10.0.0.0 to 10.255.255.255
> 172.16.0.0 to 172.31.255.255
> 192.168.0.0 to 192.168.255.255

Overview

Suppose our organization has several thousand workstations and servers. If our organization's network is not connected to the Internet, we could theoretically assign any IP addresses to our workstations and servers. However, if we intend to connect our organization's network to the Internet, we need to acquire a recognized IPv4 address as well as have that address registered so routers on the Internet know where to forward datagrams destined to the network used by our organization. Due to the shortage of IPv4 addresses, we could assign RFC 1918 addresses to our organization's workstations and servers and map or translate those addresses into a lessor number of valid IPv4 addresses. For example, we might use the 10.0.0.0 to 10.225.255.255 private IP address block to assign private Class A addresses to 2000 network devices. We might then use NAT in conjunction with a single Class C public address to map private Class A addresses to public Class C addresses on a dynamic basis. Doing so would enable one Class C network address to be used instead of eight. However, if more than 254 users simultaneously required Internet access, some user requests would either be denied or queued until a previously used Class C address becomes available for reuse.

Advantages of Use

As previously noted in this section, the use of NAT conserves scarce IPv4 address space. However, that is not the only reason to use network address translation. Two additional reasons for its use involve security and flexibility.

Concerning security, the address translation process hides hosts from direct access from the public network. Therefore, NAT makes it more difficult for outside users to determine the true structure of an organization's network in an attempt to attack the network. Another reason to consider the use of NAT concerns network flexibility. If your organization changes the service provider, you may not be able to take your network address to a new Internet Service Provider. NAT mitigates the effect of having to obtain a new network address because it allows readdressing to occur at the address translation device. This means you can continue to use your private network addresses behind the address translation device. This also means you do not have to go from host to host changing IP addresses and subnet masks, facilitating the job of a network manager or LAN administrator.

Types of Address Translation

There are three types of network address translation that certain routers and firewalls support. Those types or methods of address translation include static NAT, pooled NAT, and port-level NAT, with the latter also referred to as Port Address Translation (PAT). Although the mapping process differs with each method, their operations are similar. That is, as packets arrive at the device performing network address translation, the private source address is translated into a public address for transmission onto the Internet. In the opposite direction, inbound packets have their public IP destination address translated into their equivalent private IP destination address based on the state of an IP address-mapping table maintained by the translating device.

Static NAT

Static network address translation represents an exception to the use of NAT to economize on IPv4 address space. Under Static NAT, the address of each host on an internal network is mapped to an address on the external network as illustrated in Exhibit 1. In Exhibit 1 it was assumed that the internal network

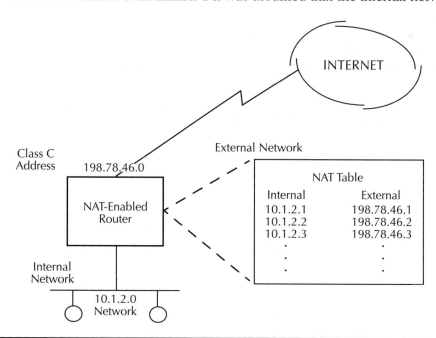

Exhibit 1.   Permanent Mapping of the Address of Each Host on an Internal Network to an Address on an External Network with Static Network Address Translation

uses the network address 10.1.2.0 while the external network uses the valid public Class C network 198.78.46.0. Because a public Class C network address is used, the private network is restricted to a maximum of 254 hosts that can access the Internet when Static NAT is used. Note that in the right portion of Exhibit 1, the NAT table shows one possible mapping between internal and external addresses. Also note that you do not have to employ sequential mapping. In fact, while doing so might be easier to administer, it also makes it easier for a hacker to examine your network structure. Although static mapping does not provide a reduction in the number of IP addresses used by an organization, it is similar to the other methods of address translation that we will discuss in that it hides the address of each workstation from the Internet. Thus, it hinders a direct attack from the external network.

A second advantage associated with Static NAT concerns the setup and operation of a translation table. Once the table is configured, no further action is required and its structure facilitates a simple table lookup process. This in turn should minimize any translation delay. However, as previously noted, a key disadvantage of Static NAT is the fact that this method of address translation does not conserve IP address space. For this reason, Static NAT is rarely used as a primary address translation method. However, it is used as a supplement to other address translation methods to alleviate certain problems we will shortly note.

Pooled NAT

A second method of network address translation is referred to as pooled NAT. Under the pooled NAT address translation technique, a group or pool of addresses is assigned to the external network for address translation purposes. Hosts on the internal network are dynamically assigned an address from the pool.

Exhibit 2 illustrates the use of a pooled NAT address translation method. In this example, it is assumed that the Class C network address 205.131.175.0

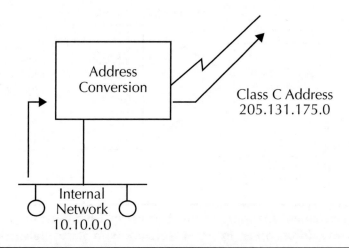

Exhibit 2.   Using a Pooled NAT Address Translation Method

is used to provide 254 unique and available public Class C addresses. In comparison, we will assume that on the internal network, we are using the RFC 1918 10.10.10.0 network address. Because this RFC 1918 address represents a private Class A network address, it is possible to use 3 bytes for host address, resulting in the ability to support $2^{24} - 2$ hosts behind the device performing the network address translation process. Obviously, most networks may only include a fraction of the number of hosts that a Class A network supports. However, you need to carefully consider the structure and potential growth of the network behind the network translation device because if the network grows, translation into a small pool of addresses may result in the inability of employees to access the Internet in a timely manner. Thus, the ability to increase the external network address pool can be an important consideration for many organizations. For example, if one Class C external network address proves to be insufficient, the ability to add one or more additional addresses is an important consideration concerning the ability of pooled NAT to respond to varying organizational requirements.

Although pooled NAT most certainly enables organizations to conserve the use of public IP addresses, its method of operation can impair the ability of certain types of applications. One example of an application a pooled NAT can inhibit is the use of Simple Network Management Protocol (SNMP).

SNMP Manager tracks devices based on the IP address of the device to be managed and an object identifier. The latter represents the address of a specific network management-related function supported by the managed object, such as the period of time it has been operating, the number of packets received in error, number of packets received, and similar data. One special type of SNMP transmission is referred to as a trap. The purpose of a trap is to enable the managed device to inform the network manager that a particular predefined condition occurred, such as an error rate or packet dropping reaching a certain level. A managed device must be configured with the IP address of the network manager for a trap to reach its intended destination. Unfortunately, this creates a problem when a pooled NAT technique is used for address translation. The reason for this is the fact that the network manager's public address under a pooled NAT technique will more than likely change over time. Thus, the trap could flow to an incorrect address behind the translation device that would not know what to do with it, perhaps resulting in an important error condition alert flowing into the great bit bucket in the sky.

Exhibit 3 illustrates how the use of a pooled NAT technique could result in the delivery of a trap to the incorrect private network host. In this example, the managed device is shown configured to transmit traps to the IP address 205.131.175.8. Unfortunately, when the trap was issued, the mapping process resulted in the assignment of a different host to that address. One possible solution to this problem is to permanently map an SNMP Manager to an IP address on the external network while all remaining internal addresses contend for the remaining addresses in the address pool. This means that if your organization needs to manage devices beyond your internal network, you might wish to consider the ability of a pooled NAT product to also support Static NAT.

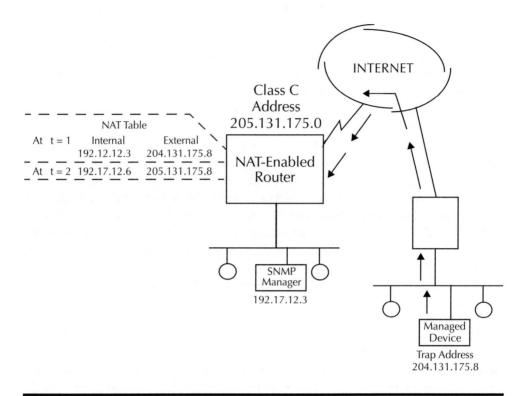

Exhibit 3. SNMP Traps Flowing to the Incorrect Host without Static IP Address Translation

Port Address Translation

A third type of network address translation results in the mapping of internal network addresses to a single IP address on the external network. While the internal to external network address mapping is static, as you might surmise, a trick is employed to enable multiple hosts behind the address translation device to correctly route responses to applicable hosts. That trick involves the assignment of different TCP and UDP port numbers to each session that uses the same static IP address on the public side of the interface.

Exhibit 4 illustrates the manner by which Port Address Translation operates. In this example all private network addresses are transmitted into the common public network address of 205.131.175.1. Port numbers used for mapping are those above 1023, providing the capability for supporting 64512 (65535 − 1023) simultaneous TCP/IP or UDP/IP connections via the utilization of a single IP address. Because address translation results in the mapping of all private network addresses to a single IP address through the use of different port numbers, this technique is also referred to as Port Address Translation (PAT). To the Internet, the use of PAT results in all traffic appearing to have originated from a single IP address.

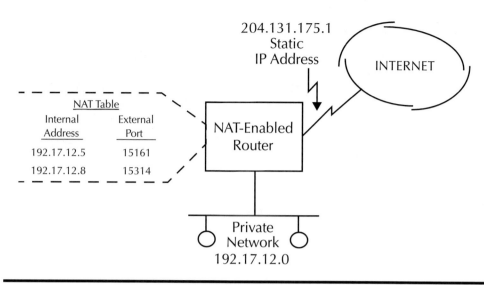

Exhibit 4. Mapping Private IP Network Addresses to a Common Public Address

Factors to Consider

There are two key factors to consider prior to implementing: (1) NAT router performance and (2) the suitability of applications.

You need to consider the effect of the translation process on equipment that will perform the translation. For example, if your organization is using a router that is performing other tasks, such as packet filtering to block pings, the NAT process could degrade router performance. This results from the fact that the router's CPU has to examine every datagram flowing from the private network onto the public network and each datagram flowing in the opposite direction. As it examines each datagram, the router must alter the IP header to change or translate addresses. In addition, the router may have to look further into the datagram to change the port number if PAT is being performed. This means that if your organization's router is heavily used, it may not be an appropriate candidate for adding NAT as an additional router function to be performed.

Application Suitability

A second factor that deserves consideration is the fact that certain Internet applications may not operate correctly when NAT is used. As noted earlier in this section, SNMP traps represent one application that requires static mapping to operate correctly. Another type of application that cannot operate under NAT are those dependent on the IP security architecture. This is because under the IP security architecture, the original headers with private network IP source addresses are digitally signed. When NAT changes the source address, it invalidates the digital signature.

Operational Example

In concluding our discussion of NAT, we will consider an example of how to configure a router to perform NAT. Because Cisco Systems has approximately 70 percent of the market for routers used on the Internet, we will examine the use of Cisco's Internetwork Operating System (IOS) to implement NAT. In doing so, assume that the Ethernet θ router port is connected to our organization's internal private network while the Serial 0 port provides a connection to the public IP-based Internet. Further assume that we will use the RFC 1918 10.0.0.0 Class A private network address internally and the public Class C network address 205.131.175.0 externally.

For static translation, assume we want to translate 10.1.2.3 to 205.131.175.3 and vice versa. In a Cisco environment, commands use the keywords *inside* and *outside* instead of *public* and *private*. The IOS commands to implement static address translation would be as follows:

```
Ip nat inside source static 10.1.2.3 205.131.175.3
Interface ethernet 0
ip address 10.1.2.1 255.255.255.0
ip nat inside
 interface serial 0
 ip address 205.131.175.1 255.255.255.0
 ip nat outside
```

In examining the previous series of IOS commands, note the Command ip nat inside source static followed by the inside (private) and outside (public) ip addresses informs the router we are performing static NAT. The two addresses inform the router of the inside and outside addresses to be translated. The next three commands assign the ip address 10.1.2.1 to the ethernet 0 interface and define the interface as being on the inside for address translation purposes. Similarly, the next three commands associate the ip address 205.131.175.1 to the serial 0 router interface and define that interface as on the outside for address translation purposes.

Continuing our examination of Cisco router commands to implement NAT, we will examine how to perform translation using a pool of addresses. In doing so, assume we want to translate addresses in the range 10.1.2.0 through 10.1.2.255 to the range 205.131.175.3 to 205.131.175.22. The following Cisco IOS statements would be required to affect the dynamic pooled NAT process previously described.

```
ip nat pool LegalPool 205.131.175.3 205.131.175.22
access-list 10 permit 10.1.2.0 0.0.0.255
ip nat inside source list 10
pool LegalPool
inside ethernet0
ip address 10.1.2.1 255.255.255.0
ip nat inside
```

```
interface serial0
ip address 205.131.175.1 255.255.255.0
ip nat outside
```

In examining the IOS commands, note that the first line set up the addresses in the address pool. The second line established a standard access list, which results in source addresses matching list 10 getting translated to addresses defined from the pool named LegalPool.

In a Cisco router environment, port address translation is referred to as port multiplexing. To initiate port multiplexing, we would add the keyword *overload* at the end of the command line *ip nat inside source*. This keyword informs the router to initiate port multiplexing with the two command lines now structured as follows:

```
Ip nat inside source list 10
Pool LegalPool overload
```

Now that we have an appreciation for the operation of network address translation, we will examine IP naming services.

IP Naming Services

There are two naming services that can be used to minimize the need to allocate IP addresses to every host on a network. One naming service, known as the Dynamic Host Configuration Protocol (DHCP), is applicable to numerous operating systems ranging from UNIX to Windows NT and Windows 2000. The second naming service examined in this section is the Windows Internet Naming Service (WINS), which as its name implies is only applicable to the Windows operating system.

DHCP

The Dynamic Host Configuration Protocol (DHCP) has a long history because its roots lie in the TCP/IP Boot Protocol (BOOTP) system that was developed to support diskless workstations on a network. BOOTP was used to provide power-on or boot-up instructions and configuration information by pointing a diskless workstation to a server from which it downloaded applicable information.

Overview

DHCP is specified in a series of four RFCs — 1533, 1534, 1541, and 1542. It is defined as a client-server protocol, with the server portion designed to provide two key services that can significantly ease the effort of network administrators: (1) providing a mechanism for many TCP/IP parameters to be defined at the server for the entire network and distributed to clients and

(2) providing for the automatic configuration of TCP/IP services on client computers. In doing so, DHCP provides another key service which is a method for allocating IP addresses to DHCP clients. In this section, we will primarily focus attention on the Microsoft DHCP service which is supported by Windows NT and Windows 2000.

Microsoft's DHCP Service

Microsoft's DHCP service represents a client-server solution for administrating the allocation of scarce IPv4 addresses. Microsoft's DHCP server includes a graphical administrative tool referred to as DHCP Manager. DHCP Manager permits users to define client configurations as well as a database for managing the assignment of client IP addresses and TCP/IP configuration parameters.

DHCP Scopes

Under DHCP, hosts are organized into groups referred to as scopes. A scope represents a logical division of hosts rather than a physical separation. You can consider a scope as an administrative grouping that is used to identify the configuration parameters for all DHCP clients grouped together on a physical subnet.

Exhibit 5 illustrates the use of the DHCP Manager's Scope menu. From this menu you can create and delete scopes. Since we want to obtain knowledge concerning the use of DHCP, we will select the create option in the pull-down menu shown in Exhibit 5.

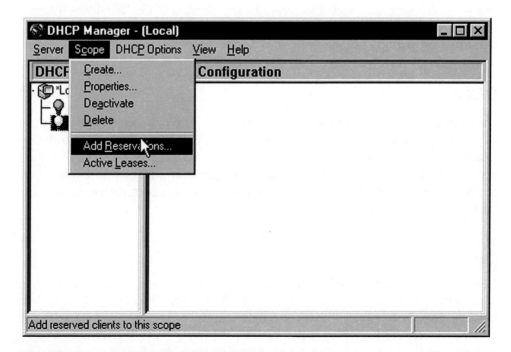

Exhibit 5. The Scope Menu in DHCP Manager

Exhibit 6 illustrates the Scope Properties dialog box after several entries were completed. Scopes must be named when they are created. In the example shown in Exhibit 6, you will note we assigned the name *barniefife* to this scope. In examining Exhibit 6, also note that we assigned the address block 205.131.175.1 through 205.131.175.254 to the address pool and did not specify any address exclusions. In the lower portion of Exhibit 6, also note that you can specify a time interval, known as a lease duration, that specifies the length of time a DHCP client can use an assigned IP address before it must renew its configuration with the DHCP server.

Options

In addition to specifying DHCP scope information, you can use DHCP Manager to define individual scope options. Those options include *deactivate* to release an IP address, *renewal* to change the renewal period for an IP address lease, and *reserve* to reserve a specific IP address for a DHCP client. The latter option provides you with the ability to reserve addresses to devices that require a fixed address, such as a Web server or a computer operating as a network firewall.

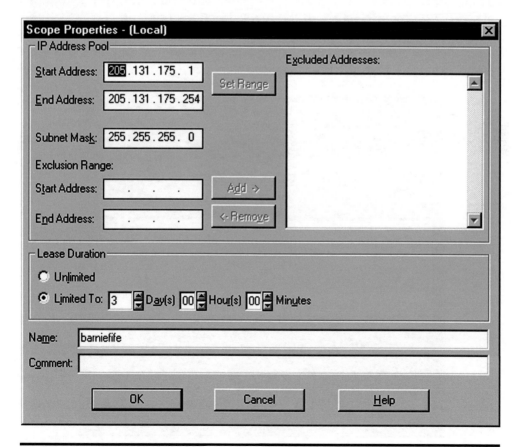

Exhibit 6. The Scope Properties Dialog Box

Viewing Activity

Once you activate DHCP, you can view leasing activity through the use of the Active Leases dialog box. Exhibit 7 illustrates an example of this dialog box. Note that in the client window shown in Exhibit 7, information concerning leases by name or IP address or data concerning reservations will be displayed, with the display based on the buttons or box checked. Because this screen was captured right after a DHCP scope was activated, no active leases are shown.

DHCP Options

DHCP supports a rich set of options that provides network managers and LAN administrators with the ability to tailor many configuration features. In a Microsoft Windows environment, you can use the DHCP Options dialog box, shown in Exhibit 8, to change one or more of a large series of predefined option values.

In examining Exhibit 8, the left window lists unused options at a particular point in time. As you select an option in that window, information about it will be displayed in the lower portion of the dialog box. You can activate an unused option by selecting it in the unused options box and clicking on the button labeled *add*. Similarly, If you have one or more active options listed in the window labeled with that name, you could remove one or more such options using the button labeled *remove*.

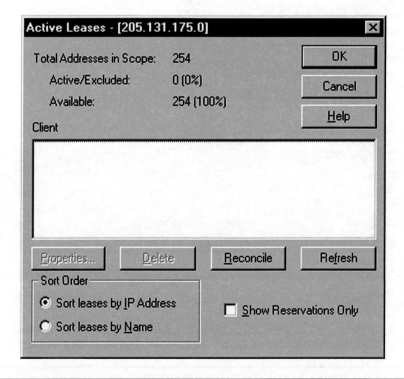

Exhibit 7.   The Active Leases Dialog Box

Microsoft follows RFC 1541 which defines default options. You can view those options by displaying the DHCP Options: Default Values dialog box that is illustrated in Exhibit 9.

While there are approximately 60 DHCP options to consider, only a small subset of 8 options has default settings. One of the default settings specifies the renewal time value that can be changed when you create a scope as

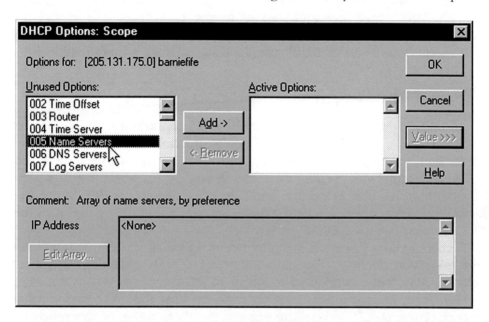

Exhibit 8. Activating Options with the DHCP Options Dialog Box (Note: The lower portion of the dialog box provides comments concerning each selected option.)

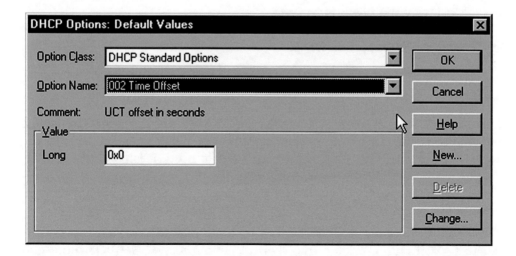

Exhibit 9. Viewing the Default Values of a Core Set of Options through the DHCP Options Dialog Box

illustrated in Exhibit 6. Other default values include specifying a list of IP addresses for NetBIOS name servers, specifying the DNS domain name that the client should use for DNS host name resolution, and specifying a list of IP addresses for DNS name servers available to the client. Concerning the latter, this illustrates the ability of network managers and LAN administrators to make a single change on the DHCP server and have all clients use a different DNS name server instead of having to visit each workstation to manually change the IP configuration. Thus, in addition to conserving IP address space, the use of DHCP can facilitate network administration.

Client Side Operations

When DHCP is running on a network, the configuration of a client resembles a trivial process. In a Windows environment, the host operator only needs to go to Start > Settings > Control Panel and click on the network icon. After TCP/IP is highlighted, clicking on a button labeled *Properties* displays a pair of radio buttons, one associated with the label *Obtain an IP Address Automatically*, which when specified informs the client that they will obtain their IP addressing from the DHCP server.

If you carefully inspect the configuration screen of a client, you will note that once you click on the button associated with obtaining an address automatically, there is no field to fill in to tell the client where the server resides. Thus, an interesting question concerns how the client knows where to contact the DHCP server. The answer to this question is the process of broadcasting. That is, the first time a DHCP client computer attempts to start on the network, it will broadcast a DHCP discovery packet using the source address 0.0.0.0 because it does not have an IP address. When the server receives the request, it selects an unleased IP address and returns that address along with additional configuration information, such as the subnet mask and default gateway the client is to use.

Now that we have an appreciation for DHCP, this chapter will conclude by turning our attention to WINS.

WINS

Microsoft's Windows Internet Name Service (WINS) was developed to use the features of DHCP for name resolution. Although WINS does not have to run with DHCP, more often than not when WINS is used, it is used with DHCP.

Overview

WINS was developed as a mechanism to resolve NetBIOS computer names, such as \\Gil and \\printer, into dynamic IP addresses. NetBIOS represents a relatively old session-level interface used by computer programs to communicate over NetBIOS-compatible transports, to include TCP/IP. Under NetBIOS, logical names were assigned to computers. While this addressing scheme made life simple for small networks that represented an isolated island in today's scheme of interconnectivity, it also precluded the use of the naming scheme to be used in a TCP/IP environment, resulting in the development of WINS to provide a translation process that converts computer names into IP addresses. In actuality WINS represents one of several methods that can be used to resolve NetBIOS computer names into IP addresses. You can use static mapping files, such as LMHOSTS, IP broadcast, or a NetBIOS name server that in a Windows NT or Windows 2000 network environment is the Microsoft WINS server.

Server Operation

Similar to DHCP, under Windows NT and Windows 2000, you would install WINS Manager on a server to support WINS.

WINS servers employ a replicated database, which contains NetBIOS computer names and IP address mappings. As clients begin to log on to the network, they transmit their computer name and obtain an IP address under a process we will review when discussing client side operations later in this section. During the log-on process, the WINS server database is updated, which provides the ability to replicate database updates among multiple WINS servers that might be required to support large networks. When multiple WINS servers are configured for use, each server is configured as a *push* or *pull* partner of at least one other WINS server, with the configuration governing the manner by which database replication occurs. Microsoft's detailed guide to WINS should be referenced if you require multiple WINS servers on a network.

Client Operations

Unlike the configuration process involved in setting up a host to use DHCP that requires a simple click on a radio button, the configuration process to support WINS is a bit more involved. To illustrate this, consider Exhibit 10, which illustrates an example of how you would enable the use of WINS on a Windows 95 workstation. In this example, note that we added the IP addresses for two WINS servers. A WINS client will perform a NetBIOS computer name-to-IP address mapping resolution process by using NetBIOS over TCP/IP (NetBT) with the first address configured. If the resolution times out, the next address will be used. Thus, while a bit more demanding than the configuration process for a host to use DHCP, the configuration of a client to use WINS is still a relatively simple process.

When a WINS-enabled computer contacts a WINS server, it will register its name using a name registration request message. The WINS server will check its database to ensure the name request is unique and respond with either a positive or negative name registration response. Assuming the registration is granted, the client must attempt to renew their registration prior to when a time-to-live value configured on the server expires. In fact, a WINS client will attempt to refresh its name registration after one-half of the time-to-live duration is expired. If the client does not receive a name refresh response

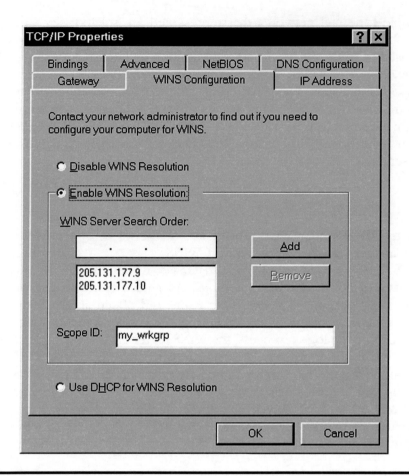

Exhibit 10.   The WINS Configuration Tab in the TCP/IP Properties Dialog Box

from the server, it will send a name refresh request to the server at 2-minute intervals. Thus, like a pest, making a nuisance of itself will hopefully satisfy the client.

Summary

WINS Manager provides a similar range of services as DHCP Manager. Through the use of WINS Manager you can specify how often a client registers its name, the interval between when an entry in the mapping table is marked as released, when it is marked as extinct, and the extinction time out which defines when an entry actually disappears from the database.

Until the introduction of Windows 2000, there were many organizations that used WINS instead of DNS. First and foremost, WINS provides direct support of NetBIOS names during the IP address resolution process which represents the key reason for its use. However, under Windows 2000, Microsoft introduced its active directory which supports a hierarchical structure similar to DNS. As users migrate to Windows 2000, we can expect the use of WINS to decrease and eventually diminish to a very low level of use.

Chapter 8

Working with IPv6

Chapter 8 will focus on a technology formerly referred to as the Internet Protocol Next Generation or Ipng. Over a considerable period of time, Ipng evolved from a mechanism designed to increase IP address space to a new IP developed to facilitate routing, security, and the addition of functionality without a requirement to again revise the protocol.

As noted earlier in this book, the near exponential growth in the use of the Internet has rapidly depleted the quantity of available IP network addresses and has resulted in such addresses becoming a precious commodity. As we turn our attention to IPv6, we will also note that its addressing capability ensures that every man, woman, and child on the planet as well as every electronic device can obtain an IPv6 address. This capability provides a mechanism to enable the development of intelligent network-based home appliances and other devices that could be managed by a service organization or the homeowner from their office. Thus, IPv6 can be considered to provide a foundation for extending the capabilities of the Internet to new applications that can be expected to arise during the new millennium.

Terminology

Under IPv6, some of the terms we grew acquainted with under IPv4 were revised. In addition, IPv6 introduces a few new terms. While we will temporarily become IPv6 *purists* and examine the terminology associated with the new protocol in this section, for reasons that will be given shortly, we will revert to IPv4 terminology in the remainder of this book.

In this section, five terms and what they mean or reference under IPv6 will be discussed:

- *Packet* — The term packet is used to reference an IPv6 header plus the following payload. If we remember the discussion of IPv4 presented earlier in this book, we used the term datagram to reference an IPv4 header plus its payload. Therefore, you might wonder why the designers of IPv6 switched the term datagram to packet. Were they seeking job security because their managers might now become confused or was there a more rational explanation for the change in nomenclature? The answer to this question resides in the fact that one innovation of IPv6 is its ability to transport traffic from other protocols than the TCP/IP protocol suite. Thus, from a theoretical basis, the term datagram would be inappropriate when the IPv6 header transports a payload that does not reside in the TCP/IP protocol suite and this resulted in the use of the term packet. However, whenever the IPv6 header transports a protocol in the TCP/IP protocol suite, such as TCP, UDP, and ICMP, the term datagram is still appropriate to use. Because I will restrict my examples to the TCP/IP protocol suite, the term datagram instead of packet will be used.
- *Node* — Under IPv6, the term node is used to represent any device that supports IPv6. Therefore, a node can represent a router, server, gateway, or client computer supporting IPv6.
- *Router* — Under IPv6, a router represents a node that forwards packets not explicitly addressed to the device.
- *Link* — A link represents a medium over which nodes communicate at the data link layer. By not defining the medium, this permits a significant degree of flexibility, enabling IPv6 to be used for mobile wireless and fixed wireless applications.
- *Neighbor* — Under IPv6, a group of nodes attached to the same link represents a neighbor to the other links.

Now that we have an appreciation of the terminology used under IPv6, we will turn our attention to the protocol by examining the IPv6 header.

The IPv6 Header

Exhibit 1 illustrates the IPv6 header. In examining the new header, you will note a considerable difference when comparing it to the IPv4 header. Although the Version field is common to both headers, once past that field, the similarity between IPv4 and IPv6 headers considerably decreases. To obtain an appreciation for the differences between the two versions of IP, let us examine each of the fields within the IPv6 header.

Version Field

Similar to the IPv4 Version field, the IPv6 Version field is 4 bits in length. As you might expect, the value of this field is set to 6 (binary 1010) for IPv6. Because the IPv6 Version field is at the beginning of the header, it provides routers and gateways with a mechanism to rapidly determine the version of the IP protocol in use.

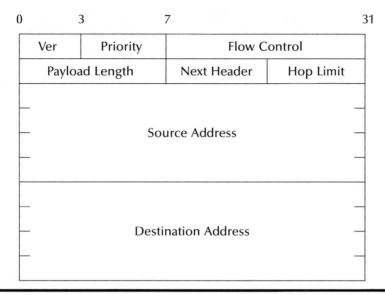

Exhibit 1. The IPv6 Header

Priority Field

The purpose of the priority field is to enable a source to identify the desired delivery priority of a datagram. While similar to the IPv4 Type of Service (TOS) field, there are distinct differences between the two. Whereas the IPv4 TOS field provides 3 bits to indicate eight levels of precedence, the IPv6 Priority field uses 4 bits to define priority for two categories of traffic. In the IPv6 Priority field, values 0 to 7 are used for traffic that backs off in response to network congestion, such as TCP segments. In comparison, values 8 to 15 are used for traffic that does not back off, such as real-time data being transported at a constant bit rate. Exhibit 2 lists the recommended priority field values for congestion controlled traffic.

In examining the entries in Exhibit 2 note that *filler* traffic represents traffic that can be transported between other traffic due to its lack of time dependence. One example of this type of traffic is NetNews. Also note that unat-

Exhibit 2. Recommended IPv6 Priority Field Values for Congestion-Controlled Traffic

Field Value	Meaning
0	Uncharacterized traffic
1	Filler traffic
2	Unattended data transfer
3	Reserved for future use
4	Attended bulk transfer
5	Reserved for future use
6	Interactive traffic
7	Internet control traffic

tended data transfer can be e-mail, while attended bulk transfer examples include FTP and HTTP applications. Last but not least, interactive traffic can include telnet, while an example of Internet control traffic would be a routing protocol, such as RIP.

Turning our attention to the second type of traffic governed by the Priority field, noncongestion controlled traffic, the lowest priority (8) should be used for datagrams that the sender is most willing to have discarded under congestion conditions, such as when a video conference is active and congesting a network. In comparison, the highest value (15) should be used for datagrams that the sender is least willing to have discarded, such as voice-over IP.

Flow Label Field

The Flow Label field is 24 bits in length. This field provides a Quality of Service (QOS) capability, which enables datagrams belonging to a particular traffic flow to be identified. Thus, a sender can use the Flow Label field in conjunction with the Priority field to convey special handling information. When the value of this field is set to zero, it indicates traffic is not associated with a flow. An example of a traffic flow would be datagrams transporting a videoconference, while HTTP and Telnet would represent examples of non-flow traffic.

Payload Length Field

The Payload Length field identifies the length of the datagram following the header. This field is 16 bits in length and supports a maximum value of 65535, which is also the maximum length of an IPv4 payload.

Next Header Field

The inclusion of a Next Header field represents a considerable improvement to the IPv6 protocol because it provides the foundation for daisy chain operations. As its name implies, the value of this field identifies the type of header following the IPv6 header. The values used for this field are the same values supported by the IPv4 Protocol field. Some examples of the Next Header field values include 1 for ICMP, 6 for TCP, 17 for UDP, and 58 for ICMPv6. Because the Next Header field provides the initial mechanism for the daisy chaining of IPv6 headers, let us examine how this is accomplished prior to resuming our examination of the fields in the IPv6 header.

Exhibit 3 illustrates an example of IPv6 header daisy chaining. In examining Exhibit 3, note that the top portion of the illustration shows IPv6 transporting UDP without any additional headers. The middle portion of Exhibit 3 indicates how header chaining can occur. In the middle portion of Exhibit 3, the IPv6 Next Header field indicates that an authentication header follows. The authentication header then notes that the next header is the

IPv6 Header Next Header = UDP	UDP Header	Data	

IPv6 Header Next Header = Authentication	Header Next Header = TCP	TCP Header	Data

IPv6 Header Next Header = Routing	Routing Header Next Header = TCP	TCP Header	Data

Exhibit 3.   IPv6 Header Daisy Chaining

TCP header. In the lower portion of Exhibit 3, the IPv6 header's Next Header field indicates that the following header is a routing header. The Next Header field in the routing header indicates that the following header is a TCP header. Note that by examining the value in the IPv6 Next Header field, a router can determine whether or not it needs to look further into a datagram. For example, when a router notes that a routing header follows the IPv6 header, it must then look further into the datagram. Because most datagrams do not require a router to look further into their contents, the ability to rapidly determine if further viewing of the contents of a datagram is necessary can significantly expedite the flow of datagrams. In addition, it also eliminates a significant amount of router processing. Thus, the use of IPv6 can be expected to extend the useful life of many routers by lowering their processing requirements.

Hop Limit Field

The Hop Limit field is 8 bits in length. The value in this field indicates the maximum number of hops a datagram can take prior to it being discarded.

The value of the Hop Limit field is set by the source and decremented by one as it flows through a router. If the field value reaches zero, a router sends the datagram to the great bit bucket in the sky. Note that under IPv4, a Time To Live (TTL) field performs a similar function to prevent datagrams from forever wandering the Internet or a private intranet.

Source Address Field

The IPv6 source address field increased to 128 bits from IPv4's 32-bit structure. This field is similar to the IPv4 source address field in that it identifies the originator of a datagram. The next section in this chapter will examine the structure of IPv6 addresses in detail when we discuss the address architecture of IPv6.

Destination Address Field

Similar to the Source Address field, the IPv6 Destination Address field is 128 bits in length. This field identifies the intended recipient of a datagram.

Address Architecture

IPv6 is based on the same architecture used in IPv4, resulting in each network interface requiring a distinct IP address. The key differences between IPv6 and IPv4 with respect to addresses are the manner by which an interface can be identified and the size and composition of the address. Under IPv6, an interface can be identified by several addresses to facilitate routing and management. In comparison, under IPv4, an interface can only be assigned one address. Concerning address size, IPv6 uses 128 bits or 96 more bits than an IPv4 address.

Address Types

IPv6 addresses include unicast and multicast, which were included in IPv4. In addition, IPv6 adds a new address category known as anycast. Although an anycast address identifies a group of stations similar to a multicast address, a packet with an anycast address is delivered to only one station, the nearest member of the group. The use of anycast addressing can be expected to facilitate network restructuring while minimizing the amount of configuration changes required to support a new network structure. This is because you can use an anycast address to reference a group of routers and the alteration of a network when stations use anycast addressing would enable them to continue to access the nearest router without a user having to change the address configuration of their workstation.

Address Notation

Since IPv6 addresses consist of 128 bits, a mechanism is required to facilitate their entry as configuration data. The mechanism used is to replace those bits by the use of eight 16-bit integers separated by colons, with each integer represented by 4 hexadecimal digits. For example, the following represents the entry of a 128-bit Ipv6 address using hex notation.

```
6ACD:00001:00FC:B10C:0001:0000:0000:001A
```

To facilitate the entry of IPv6 addresses, you can skip leading zeros in each hexadecimal component. That is, write 1 instead of 0001 and 0 instead of 0000. Thus, this ability to suppress zeros in each hexadecimal component would reduce the previous network address to:

```
6ACD:1:FC:B10C:1:0:0:1A
```

Under IPv6, a second method of address simplification was introduced — the double-colon (::) convention. Inside an address, a set of consecutive null 16-bit numbers can be replaced by two colons (::). Thus, the previously reduced IP address could be further reduced as:

```
6ACD:1:FC:B10C:1::1A
```

It is important to note that the double-colon convention can only be used once inside addresses. This is because the reconstruction of the address requires the number of integer fields in the address to be subtracted from eight to determine the number of consecutive fields of zero value the double-colon represents. Otherwise, the use of two or more double-colons would create ambiguity that would not allow the address to be correctly reconstructed.

Address Allocation

The use of a 128-bit address space provides a high degree of address assignment flexibility beyond that available under IPv4. IPv6 addressing enables Internet Service Providers to be identified as well as includes the ability to identify local and global multicast addresses, private site addresses for use within an organization, hierarchical geographical global unicast addresses, and other types of addresses. Exhibit 4 lists the initial allocation of address space under IPv6.

The Internet Assigned Numbers Authority (IANA) was assigned the task to distribute portions of IPv6 address space to regional registries around the world, such as the InterNIC in North America, RIPE in Europe, and APNIC in Asia. To illustrate the planned use of IPv6 addresses, we will turn our attention to what will be the most common type of IPv6 address — the provider based address.

Provider-Based Addresses

The first official distribution of IPv6 addresses will be accomplished through the use of provider-based addresses. Based on the initial allocation of IPv6 addresses as shown in Exhibit 4, each provider-based address will have the 3-bit prefix 010. That prefix will be followed by fields that identify the registry that allocated the address, the service provider, and the subscriber. The latter field actually consists of three subfields: (1) a subscriber ID that can represent an organization, (2) a variable network and interface identification fields used in a similar manner to IPv4 network, and (3) host fields. Exhibit 5 illustrates the initial structure for a provider-based address.

Special Addresses

Under IPv6 there are four special types of unicast addresses that were defined. This section will briefly examine each address.

Exhibit 4. IPv6 Address Space Allocation

Allocation	Fraction of Prefix (binary)	Address Space
Reserved	0000 0000	1/256
Unassigned	0000 0001	1/256
Reserved for NSAP allocation	0000 001	1/128
Reserved for IPX allocation	0000 010	1/128
Unassigned	0000 011	1/128
Unassigned	0000 1	1/32
Unassigned	0001	1/16
Unassigned	001	1/8
Provider-based Unicast Address	010	1/8
Unassigned	011	1/8
Reserved for geographic-based Unicast Address	100	1/8
Unassigned	101	1/8
Unassigned	110	1/8
Unassigned	1110	1/16
Unassigned	1111 0	1/32
Unassigned	1111 10	1/64
Unassigned	1111 110	1/128
Unassigned	1111 1110 0	1/512
Link-local use addresses	1111 1110 10	1/1024
Site-local use addresses	1111 1110 11	1/1024
Multicast addresses	1111 1111	1/256

Prefix	Registry ID	Provider ID	Subscriber ID	Subnet ID	Station ID

Legend:

Prefix	3 bits set to 010
Registry	5 bits identifies organization that allocated the address
Provider	24 bits with 16 used to identify ISP and 8 used for future extensions
Subscriber	32 bits with 24 used to identify the subscriber and 8 used for extension
Subnet	16 bits to identify the subnetwork
Station	48 bits to identify the station

Exhibit 5. Provider-Based Address Structure

Unspecified Address

The all-0s address under IPv6 is referred to as an unspecified address. The composition of that address in its reduced form is shown below:

```
0:0:0:0:0:0:0:0
```

The primary use of an unspecified address is by a device that has yet to learn its address. In this situation, the device will transmit a datagram using a source address of all-0s.

Version 6 Loopback

If you remember our discussion of special IPv4 addresses, we noted the use of the 127 Class A network for loopback. Under IPv6, there is only one loopback address whose composition is noted below:

```
0:0:0:0:0:0:0:1
```

Version 4 Addresses

In a mixed IPv4 and IPv6 environment, devices that do not support IPv6 will be mapped to version 6 addresses using the following form:

```
0:0:0:0:0:FFFF:w.x.y.z
```

Here w.x.y.z represents the original IPv4 address. Thus, IPv4 addresses will be transported as IPv6 addresses through the use of the IPv6 version 4 address format. This means that an organization with a large number of workstations and servers connected to the Internet only has to upgrade their router to support IPv6 addressing when IPv6 is deployed. Then, they can gradually upgrade their network on a device-by-device basis to obtain an orderly migration to IPv6.

Version 6 Addresses Interfacing V4 Networks

Another special IPv6 address that warrants attention is the address format used by Version 6 devices communicating with one another across an Ipv4 network. Because IPv6 must be transported as IPv4 datagrams, the term IPv4 tunneling is used to reference this method.

To support IPv4 tunneling, each IPv6 node attached to an IPv4 network is assigned an IPv4 address at the interface to the IPv4 network. The IPv6 address is then mapped to a special IPv4 compatible IPv6 address format as indicated below:

```
0:0:0:0:0:0:w.x.y.z
```

In examining the preceding format, note that the first 96 bits in the IPv6 address are set to zero, in effect creating an IPv4 address for tunneling purposes.

Although IPv6 is being used on an experimental portion of the Internet, its anticipated movement into a production environment was delayed due to the more efficient use of existing IPv4 addresses. This occurred via network address translation and the use of DHCP. While the use of IPv6 is less pressing than thought just a few year ago, no matter how efficient the allocation of the remaining IPv4 addresses become, it is a known fact that within the next few years all available addresses will be used. Prior to that time, we can expect a migration to IPv6 to occur.

Chapter 9

Network Utility Tools

In this concluding chapter, attention will turn to the operation and utilization of a core series of network utility tools. The tools we will examine include Ping, traceroute, Pathping, nslookup, and ipconfig. While Ping, traceroute, and nslookup are applicable to different operating systems that support a TCP/IP protocol stack, pathping represents a proprietary tool introduced under Windows 2000 and is currently restricted to that operating system.

The network utility tools examined in this chapter were developed to provide TCP/IP users with a mechanism to isolate problems, determine the latency through a network, examine the manner by which datagrams are routed through a network, and determine information about a host name or IP address.

Ping

The Ping utility program is included in every operating system known to me to support the TCP/IP protocol stack. Although many persons associate the name of the program to radar or sonar operations because a ping returns information from a target or destination, another story or legend traces the name to the term packet Internet Grouper. In actuality, Ping was created by Mike Muuss of the U.S. Army Research Laboratory in December 1983. Mr. Muuss required a mechanism to determine if various network systems were operational. His work initially was commonly used as a verb: *Ping server/gateway/router to see if it's up.* Ping is now an Internet standard and is defined in RFC 792.

Regardless of the derivation of the name of the program or whether you refer to it as a verb or noun, Ping operates by transmitting an ICMP echo-request datagram to a target computer. If the destination machine is operational and supports the TCP/IP protocol stack, it responds with an ICMP echo-response.

Format

The general format of the Ping utility program in a Windows operating environment is illustrated in Exhibit 1. In its basic format, a person commonly enters the command Ping followed by a host name, such as pingwww.yale.edu to ping the Web server at Yale University. When you use Ping without specifying any operations, Windows will automatically generate a sequence of 4 echo-request packets, each 32 bytes in length. Through the use of different Ping options, you can tailor the use of this utility program. Thus, let us examine some of the options supported by the Microsoft version of Ping. Once this is accomplished, we will examine use of this utility program.

Options

This section will address the options supported by the Ping utility program.

-t Option

The -t option results in the continuous pinging of the target or destination device. Many times you may need to determine the average response time from one host to another through a network. To do so you can either sit in front of your computer and issue one Ping after another and record each round-trip delay summary and compute the average response time or you can use a little trick. That trick actually involves two functions. First, issue the Ping with the -t option and pipe the results into a file. For example, under

```
MS-DOS Prompt                                              _ 5 x

Auto    ▼  □ ▣ ▣ ▣ ▣ ▣ A

C:\WINDOWS>ping

Usage: ping [-t] [-a] [-n count] [-l size] [-f] [-i TTL] [-v TOS]
            [-r count] [-s count] [[-j host-list] | [-k host-list]]
            [-w timeout] destination-list

Options:
    -t              Ping the specified host until stopped.
                    To see statistics and continue - type Control-Break;
                    To stop - type Control-C.
    -a              Resolve addresses to hostnames.
    -n count        Number of echo requests to send.
    -l size         Send buffer size.
    -f              Set Don't Fragment flag in packet.
    -i TTL          Time To Live.
    -v TOS          Type Of Service.
    -r count        Record route for count hops.
    -s count        Timestamp for count hops.
    -j host-list    Loose source route along host-list.
    -k host-list    Strict source route along host-list.
    -w timeout      Timeout in milliseconds to wait for each reply.

C:\WINDOWS>
```

Exhibit 1. The Microsoft Version of Ping

Windows you would use the command prompt to enter the following command to continuously Ping the Yale University server and pipe the responses into a file named most appropriately pingresponse.

```
Ping -t www.yale.edu.>pingresponse
```

After what you feel is a sufficient period of time, enter the multi-key Ctrl-C. You would then import the file into a spreadsheet program, such as Excel or Lotus 123, and manipulate the entries to determine the statistics you require.

-a Option

The -a option results in an IP address being resolved into a hostname. Note that this is the reverse of the default, since Ping automatically resolves a hostname into an IP address. Concerning the default, a few words of caution are required if you intend to use Ping to determine the round-trip delay as a basis for implementing application. If you use a host name in the command that was not previously resolved into an IP address, a DNS name resolution process will occur. Depending on where in the DNS hierarchy the resolution process obtains the applicable IP address, the time required for the first Ping can be delayed, whereas subsequent Pings are not. If you limit your Ping operation to typing the command a single time, the first-round trip delay may be higher than the other three times because Microsoft generates a sequence of four Pings by default. Therefore, in this situation, 25 percent of the response could be in error. To remove this potential error, simply issue the Ping command two or more times and discard the first set of results because only that group would be adversely affected by the name resolution process.

-n Option

Continuing our examination of Microsoft Ping options, the -n option permits us to change the number of echo requests that will be transmitted when a Ping command is entered. For example, to Ping the Yale University Web server a dozen times, we would use the command as follows:

```
Ping -n 12 www.yale.edu
```

-l Option

Through the use of the -l option, you can adjust the length of the datagrams transmitted.

-f and -i Options

The -f option sets the do not fragment flag in the IP header while the -i option sets the Time to Live (TTL) field value in the IP header. Concerning the latter, you can set the TTL field to a specific value to determine if routers are discarding traffic correctly. For example, if a datagram needs to flow through five routers on its path from source to destination, you can use Ping with different TTL values to determine if each router on the path is correctly sending packets to the great bit bucket in the sky.

-v Option

The -v option can be used to set the Type of Service (TOS) field to a specific value. You can use this option as a mechanism to test the priority queuing configuration of a router. For example, assume you configured a router to expedite the flow of datagrams based on the setting of the TOS field in the IP header. You can then transmit multiple data streams through the router along with Pings with a predefined TOS setting and analyze the results of traffic flowing through the router to determine if the queuing method configured produces the required results.

-r Option

The -r option when set results in the route of the outbound datagram being recorded and returned in the response. A minimum count of one and maximum count of nine hosts must be specified when this option is used.

-s Option

The -s option results in a timestamp operation being performed for the number of hops specified when the -r option is used.

-j Option

The -j option provides you with the ability to route datagrams via the list of hosts specified by host-list. Consecutive hosts can be separated by intermediate gateways which results in the term *loose source route* shown in Exhibit 1. The maximum number of hosts in the host-list is nine.

-k Option

The opposite of a loose source route is a strict source route. Thus, the -k option provides you with a mechanism to specify the route that datagrams should explicitly follow. Because consecutive hosts cannot be separated by intermediate gateways, the name *strict source route* is used for this option. Similar to the -j option, a maximum of nine hosts can be specified.

-w Option

The -w option provides a mechanism to set a timeout interval in milliseconds for each reply. The default timeout is 3000 ms or 3 s. When a timeout occurs, the message *Request timed out* will be displayed.

Operational Examples

Exhibit 2 illustrates the use of Ping on the Web server at Yale University. Note that the command line does not include the use of any options, resulting in four echo-request ICMP messages being transmitted. Although three echo-responses were received, the fourth echo-response was not received within the timeout period, resulting in the message *Request time out* shown after the three lines beginning with the common term *Reply from*.

In examining the use of the Ping utility in Exhibit 2, note that Ping obtains and displays the IP address of the destination host name. Also note that the default results in 32 bytes being transmitted and received. At the bottom of Exhibit 2, you will note that Microsoft's implementation of Ping provides a summary of packets sent, packets received, packets lost, and minimum, maximum, and average round-trip delay times.

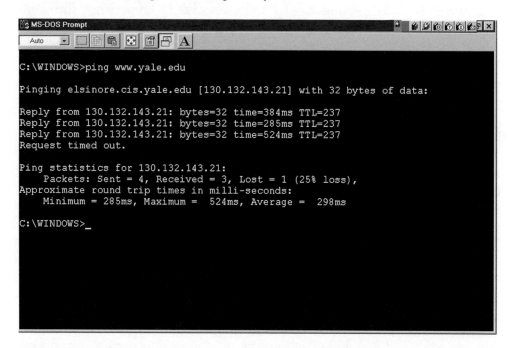

Exhibit 2. Using Ping to Determine the Round-Trip Delay to the Yale University Web Server

Utilization

One of the key uses of Ping is to ascertain if a destination is operational. However, another use of this utility is to determine if a path to the destination is available. Thus, if you receive a response to a Ping, you know that a path to the destination is available and the destination is operational and the TCP/IP protocol stack at the destination is also operational. Due to the preceding, Ping is commonly used to test equipment installed on a TCP/IP network, i.e., after you configure the TCP/IP protocol stack on a workstation, you would use Ping to ping another station on the network. Upon receipt of one or more replies, you would then know the TCP/IP protocol stack and the network connection of the workstation are operating correctly.

A second common use of Ping is to determine the one-way delay or latency through a network. Because Ping measures the round-trip delay, half of that delay represents the one-way delay from source to destination. If you anticipate the addition of a real-time application, such as voice over IP (VoIP) via the Internet or a private intranet, you should consider using Ping prior to implementing the application because Ping can inform you as to whether or not the application is suitable for use. A VoIP application cannot tolerate a delay greater than approximately 250 ms. Therefore, when the delay exceeds that time, it becomes difficult for persons having a conversation to know when the other party actually stopped talking, resulting in both parties frequently beginning to talk at the same time. Then use of the term *over* is required as a mechanism for one person to tell the other that it is their turn to talk. Obviously, this situation would not represent a viable mechanism for a business call.

Using Multiple Options

To conclude examination of the operation of Ping, we will discuss the use of multiple command options. Exhibit 3 illustrates the use of the -n and -l options on one command line. In this example, the -n option was suffixed with a value of 8 to generate eight echo requests, which in this example resulted in eight responses being received. The second option used in the Ping command line shown in Exhibit 3 is the -l option. This option was used with a value of 64 as a mechanism to generate 64-byte datagrams. Note the lower portion of the display provides summary statistics for the 8 pings, indicating that the minimum, maximum, and average round-trip delays were 189 ms, 262 ms, and 204 ms, respectively.

For a second example of the use of multiple options in a Ping command line, we will use the -n and -r options. As previously discussed, the -n option governs the number of pings generated while the -r option results in the route of up to nine hops being noted. In the example shown in Exhibit 4, I set a value of 1 for the -n option only because I wanted the results to fit on my display for illustrative purposes. The use of the -r option results in the display of nine nodes which is the maximum value for this option. From the two multioption examples, it should be obvious that it is relatively easy to tailor the use of Ping options to satisfy your particular requirements.

```
MS MS-DOS Prompt                                                      _ 8 x
 Auto    ▼  □ 🗐 🖺 🔯 🗗 🗗  A

C:\WINDOWS>ping -n 8 -l 64 www.yale.edu

Pinging elsinore.cis.yale.edu [130.132.143.21] with 64 bytes of data:

Reply from 130.132.143.21: bytes=64 time=262ms TTL=237
Reply from 130.132.143.21: bytes=64 time=204ms TTL=237
Reply from 130.132.143.21: bytes=64 time=198ms TTL=237
Reply from 130.132.143.21: bytes=64 time=192ms TTL=237
Reply from 130.132.143.21: bytes=64 time=196ms TTL=237
Reply from 130.132.143.21: bytes=64 time=192ms TTL=237
Reply from 130.132.143.21: bytes=64 time=204ms TTL=237
Reply from 130.132.143.21: bytes=64 time=189ms TTL=237

Ping statistics for 130.132.143.21:
    Packets: Sent = 8, Received = 8, Lost = 0 (0% loss),
Approximate round trip times in milli-seconds:
    Minimum = 189ms, Maximum =  262ms, Average =  204ms

C:\WINDOWS>
```

Exhibit 3.   Using the -n and -l Options in a Ping Command Line

```
MS MS-DOS Prompt                                                      _ 8 x
 Auto    ▼  □ 🗐 🖺 🔯 🗗 🗗  A

C:\WINDOWS>ping -n 1 -r 9 www.yale.edu

Pinging elsinore.cis.yale.edu [130.132.143.21] with 32 bytes of data:

Reply from 130.132.143.21: bytes=32 time=217ms TTL=237
    Route: 12.122.253.250 ->
           12.122.5.134 ->
           12.122.2.130 ->
           12.123.20.250 ->
           205.171.1.30 ->
           205.171.21.106 ->
           205.171.5.243 ->
           205.171.24.6 ->
           205.171.5.235

Ping statistics for 130.132.143.21:
    Packets: Sent = 1, Received = 1, Lost = 0 (0% loss),
Approximate round trip times in milli-seconds:
    Minimum = 217ms, Maximum =  217ms, Average =  217ms

C:\WINDOWS>_
```

Exhibit 4. Using the -n and -r Options in a Ping Command Line

PingMePlease

Although I am normally hesitant about referencing Web URLs because many locations disappear by the time a book is published, I will make an exception. In surfing the Web, I located a most interesting Web site at http://www.ping-meplease.com. This home page is illustrated in Exhibit 5. What is most interesting about this Web site is the fact that it provides a gateway to a number of locations around the globe that you can access to obtain pings from those locations. In addition, this Web site also provides links to other utility programs on the Web, such as traceroute that can be used to trace a route from different destinations back to your location. Thus, you can use this site as a learning tool about the structure of the Internet.

Tracert

Earlier in this chapter, I mentioned we would cover the traceroute utility program. Therefore, you may be a bit confused by the heading of this section. Let me digress a bit and explain. Tracert is the name Microsoft uses under Windows to reference the traceroute program. If you remember the good old days of the disk operating system (DOS), you should also remember that that operating system was limited to supporting filenames of eight characters or less. Because *traceroute* consists of ten characters, in the days of DOS,

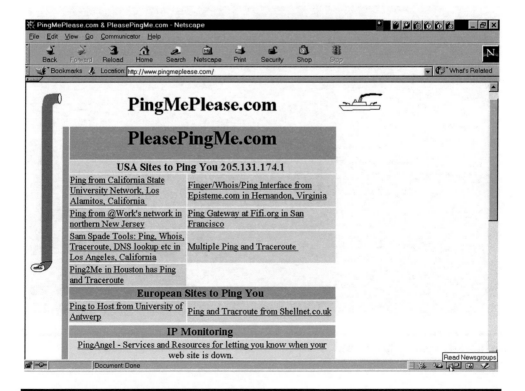

Exhibit 5.   The PingMePlease Web Page

Microsoft could not directly support the full name, resulting in the abbreviation tracert being used for the name of this utility program.

Overview

Exhibit 6 illustrates the entry of the command name without additional information to generate a help screen. The help screen lists the manner by which the tracert command is used to include its options. In examining Exhibit 6, note that tracert only has four options. You should also note that the term *target-name* references the destination for which you wish to obtain information about the route to that location. That destination, as we will shortly note, can be expressed either as a host name or as an IP address. Thus, to trace the route to the Yale University Web server, we could enter the tracert command as:

```
tracert www.yale.edu or tracert 130.132.143.21
```

where the second use of the command requires us to know the IP address of the destination.

Command Options

As previously noted, tracert has four options.

-d Option

The -d option is used to prevent the resolution of IP addresses to host names.

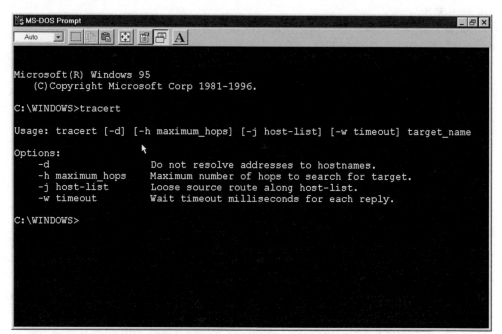

Exhibit 6. Entering Tracert without Any Parameters

-h Option

The second option, -h, permits you to specify a maximum number of hops during a search of a route to the destination specified as a target name in the command. The default value is 30 hops which should be sufficient for almost every conceivable use of tracert. However, if you need more hops to reach a destination, you can easily use the -h option. For example, assume you could not reach the Yale University Web server with the default value of 30 hops. To change the default setting to 40 hops, you would enter the following command line:

```
Tracert -h 40 www.yale.edu
```

-j Option

Continuing our examination of tracert options, the -j option permits you to specify a loose source route by listing a series of host names.

-w Option

Last but not least, the -w option permits you to specify a timeout value in milliseconds for each reply. As we will shortly note when examining the use of tracert, this utility program will discover a route by transmitting a sequence of three datagrams at a time. Thus, if it requires 30 hops to determine a route to a destination, tracert will generate 90 datagrams. You can then set a timeout threshold for each datagram by using the -w option followed by a value in milliseconds.

Now that we have an appreciation for the options supported by tracert, we will discuss its operation. In doing so, we will first discuss how tracert operates and then examine the use of the command to trace the route from my computer to the Yale University Web server.

Operation

As noted during an earlier discussion of the IP header, the Time-To-Live (TTL) field value is decremented by 1 as a datagram flows through a router. If the value reaches zero, the router sends the datagram to the great bit bucket in the sky. In addition, the router will return an ICMP Time Exceeded message to the originator of the datagram.

Recognizing the manner by which the TTL field functions as a hop counter, the designer of tracert developed the program to send a sequence of datagrams with varying TTL field values, i.e., the tracert program first sets the value of the TTL field in the IP header to 1. This results in the first router along the path to the destination setting the field value to zero, discarding the datagram, and returning an ICMP Time Exceeded message. Thus, the first router along the path to the destination has now been determined.

The tracert program next increments the value of the TTL field in the IP header by 1 and transmits another sequence of three datagrams. The program will continue incrementing the value of the TTL field in the IP header by 1 until one of two conditions occurs — (1) the target responds or (2) the maximum TTL value is reached. By examining the ICMP Time Exceeded messages returned by each router along the path to the destination, the tracert program can determine the route to the destination.

Now that we have an appreciation for the details concerning how the program learns the route to the destination, we need to consider one more item concerning routers prior to moving on: the fact that routers can be configured to not return an ICMP Time Exceeded message. This fact can explain why some routers along the path will appear invisible to the program regardless of the timeout value you use. When a router becomes invisible to tracert, the program's query will timeout and an asterisk (*) will be displayed instead of the round-trip delay.

Utilization

Exhibit 7 illustrates the use of the tracert program to trace the route to the Web server at Yale University. As you will see from examining Exhibit 7, the route from my computer to the Yale University Web server consists of 15 hops. Also note that the fifteenth hop is not a router, but the target or destination host. Thus, in actuality, the tracert program always shows one more hop than necessary when it discovers a path to a destination.

```
MS-DOS Prompt                                                    _ 🗗 ✕
Auto    ▾  ☐ 🖺 🖺  🖸 🖆 🖫 A

C:\WINDOWS>tracert www.yale.edu

Tracing route to elsinore.cis.yale.edu [130.132.143.21]
over a maximum of 30 hops:

  1   1019 ms   1388 ms    851 ms   12.92.13.17
  2    180 ms    226 ms    411 ms   gbr1-p54.ormfl.ip.att.net [12.122.253.253]
  3    326 ms    165 ms    562 ms   gbr4-p70.ormfl.ip.att.net [12.122.5.125]
  4    443 ms    166 ms    361 ms   gbr3-p10.attga.ip.att.net [12.122.2.129]
  5    176 ms    442 ms    166 ms   ggr1-p360.attga.ip.att.net [12.123.20.249]
  6    183 ms    191 ms    183 ms   atl-brdr-03.inet.qwest.net [205.171.1.29]
  7    180 ms    289 ms    179 ms   atl-core-03.inet.qwest.net [205.171.21.105]
  8    176 ms    190 ms    181 ms   wdc-core-03.inet.qwest.net [205.171.5.241]
  9    182 ms    181 ms    183 ms   wdc-core-02.inet.qwest.net [205.171.24.5]
 10    175 ms    182 ms    328 ms   jfk-core-01.inet.qwest.net [205.171.5.233]
 11    356 ms    188 ms    182 ms   bos-core-02.inet.qwest.net [205.171.8.17]
 12    185 ms    188 ms    183 ms   205.171.28.30
 13    214 ms    189 ms    188 ms   63.145.0.14
 14    406 ms    366 ms    313 ms   sloth.net.yale.edu [130.132.1.17]
 15    499 ms    469 ms    207 ms   elsinore.cis.yale.edu [130.132.143.21]

Trace complete.

C:\WINDOWS>
```

Exhibit 7. Tracing the Route from the Author's Computer to the Yale University Web Server

Router Identification

If you carefully examine the results of using the tracert program shown in Exhibit 7, you will also see that some routers include a description in addition to their IP address in an ICMP Time Exceeded message returned to the source computer issuing the tracert sequence of IP datagrams. If you look carefully at the description of the routers for hops 2 through 5, you will see they are operated by AT&T, which indicates that those routers reside on the AT&T network portion of the Internet. Similarly, the description of the routers for hops 6 through 12 indicates they are on the Quest Communications network. If you look further into the Quest router descriptions, you will see the abbreviations *abl*, *wdc*, *jfk*, and *bos* as prefixes to the router description. Those prefixes indicate where the router resides, providing additional information concerning how datagrams flow from source to destination.

Bottleneck Determination

The use of tracert provides a mechanism to trace the route to a specific destination. In doing so, the program displays information about each hop on the path to the destination. In addition, the program determines the round-trip delay to each hop, permitting you to see where bottlenecks may reside along the path from source to destination. This means that if your organization is considering implementing a time-dependent application, you can isolate bottlenecks. This provides you with the ability to either replace routers or rearrange your network structure if your organization operates an intranet or to contact your Internet Service Provider (ISP) if the problem resides on the Internet.

Now that we have an appreciation for tracert, we will turn our attention to a relatively new utility program that is currently only supported under Windows 2000. That program is called pathping.

Pathping

You may view Pathping as a route-tracing utility program that combines the features of Ping and tracert programs as well as adds information that neither of the previously described programs provides. Because of this, some persons refer to pathping as Ping on steroids.

Overview

Similar to Ping and tracert, you can display a list of pathping options by entering the command name by itself as shown in Exhibit 8. However, prior to examining program options, a few words about the location of the command prompt under Windows 2000 are in order. Under Windows 2000, access to the command prompt was moved under Programs > Accessories. This may

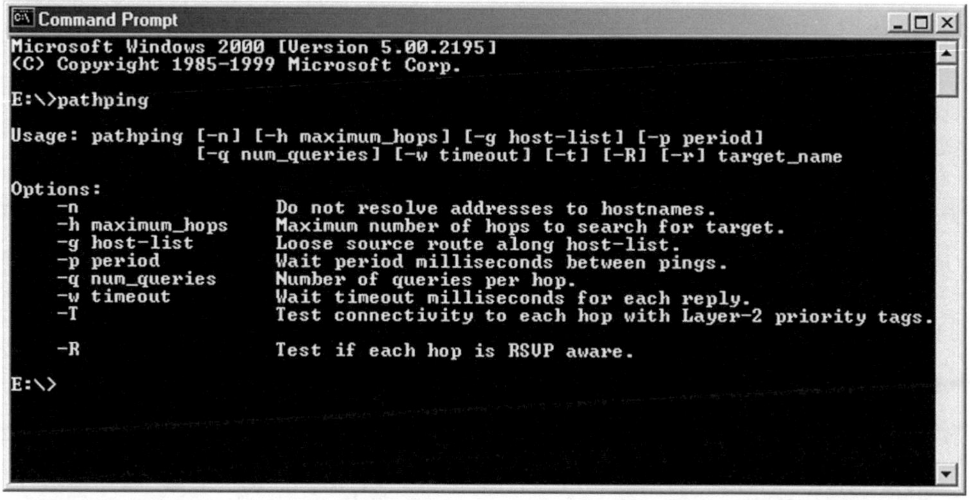

```
💻 Command Prompt                                                    _ 🗆 ✕
Microsoft Windows 2000 [Version 5.00.2195]
(C) Copyright 1985-1999 Microsoft Corp.

E:\>pathping

Usage: pathping [-n] [-h maximum_hops] [-g host-list] [-p period]
                [-q num_queries] [-w timeout] [-t] [-R] [-r] target_name

Options:
    -n                   Do not resolve addresses to hostnames.
    -h maximum_hops      Maximum number of hops to search for target.
    -g host-list         Loose source route along host-list.
    -p period            Wait period milliseconds between pings.
    -q num_queries       Number of queries per hop.
    -w timeout           Wait timeout milliseconds for each reply.
    -T                   Test connectivity to each hop with Layer-2 priority tags.

    -R                   Test if each hop is RSVP aware.

E:\>
```

Exhibit 8.   Results of Entering the Command Name Pathping by Itself

explain why some persons familiar with the use of the command prompt under other versions of Windows appear to scratch their head for a few minutes when working with this newer version of Windows.

Now that we know where the command prompt resides under Windows 2000, we will discuss pathping command options.

Options

Pathping doubles the number of tracert options, providing users with a total of eight options to consider. Similar to Ping and tracert, you can simply enter the command with a target name, resulting in default values being used. For example, returning to the Yale University Web server, we can use the pathping command as follows:

```
Pathping www.yale.edu.
```

-n Option

The first option shown in Exhibit 8, -n, results in the prevention of the resolution of IP addresses into host names. If you do not select this option, the program will automatically perform a reverse name resolution which adds time to the pathping command execution process.

-h Option

The -h option is used to specify the number of hops to search for the target. Similar to Ping and tracert, the default value is 30 hops.

-g Option

The -g option provides you with the ability to specify a loose source route. This means the hosts specified can be separated by intermediate gateways.

-p Option

The -p option provides you with the ability to specify the time or period to wait between successive pings. The default value is 250 ms or a quarter of a second.

-q Option

The -q option is used to specify the number of queries transmitted to each router along the route to the destination. The default value for pathping is 100, which is significantly greater than the use of four queries under Ping and three queries per router under tracert. While the default value of 100 is a significant improvement when you require a statistical average with a small standard deviation, it is important to note that the -p and -q options along with the number of hops required to reach a target can result in an extremely long time for the program to fully execute. For example, if the number of hops to reach the target is 20 and the default value of the -q option of 100 is retained, the program will generate 20 × 100 or 2000 queries. If you set the value of the -p option to 12 to wait 3 seconds between pings, the total duration becomes 2000 queries × 3 seconds per query or 6000 seconds which is 100 minutes. Thus, you should carefully consider the values for the -h, -p, and -q options as an entity.

-T Option

The -T option results in the attachment of a Layer 2 (Media Access Control) priority tag to each ping packet transmitted to each of the network devices along the route to the destination. You would use this option as a mechanism to identify network devices that do not support or do not have Layer 2 priority enabled. Unlike preceding options that can be upper or lower case, the -T option must be capitalized.

-R Option

The last option supported by pathping is the -R option. This option can be used to determine whether or not each device along the route to the destination supports the Resource Reservation Protocol (RSVP). This option enables you to determine if it is possible to obtain reserved bandwidth between source and destination to support an application requiring a Quality of Service (QOS) function. Similar to the -T option, the -R option must also be capitalized.

Now that we have reviewed pathping options, we will discuss its operation.

Operation

Exhibit 9 illustrates the initial execution of the pathping command using the Yale University Web server as the destination address in the command line. If you carefully examine the use of pathping shown in Exhibit 9, it might appear to resemble a stripped-down version of tracert. However, before jumping to this conclusion, note the bottom line in the illustration that displays the message *computing statistics for 325 seconds*. The reason you will have to wait 5 minutes and 25 seconds for the display of pathping statistics is due to the manner by which the program operates. Because the default operation of the program results in 100 pings per hop at intervals of 250 ms, this equates to a duration of 25 seconds (100 × 1/4 second) per router hop. Because 13 hops were required to reach the target, this results in a waiting time of 13 hops × 25 seconds per hop or 325 seconds.

We will continue our examination of the use of pathping by focusing attention on statistics generated by the program. Exhibit 10 provides a display of the continuation of the program.

If you examine Exhibit 10, note that the column labeled *RTT* defines the average round-trip time for 100 ICMP datagrams. The next column, which is labeled *Lost/Sent*, indicates the number, if any, and percent of datagrams dropped when sent to a specific hop. The following column provides similar data on a link basis while the last column provides information about the hop address similar to tracert.

Now that we have examined pathping, we will move on to the use of nslookup.

```
MS MS-DOS Prompt - pathping www.yale.edu                        _ | 8 | X |
Microsoft Windows 2000 [Version 5.00.2195]
(C) Copyright 1985-1999 Microsoft Corp.

C:\WINDOWS>pathping www.yale.edu

Tracing route to elsinore.cis.yale.edu [130.132.143.21]
over a maximum of 30 hops:
  0  passed [166.72.216.233]
  1  204.146.251.94
  2  atla1ar2-5-0-18.ga.us.prserv.net [165.87.132.178]
  3  atla1br2-11-0-0.ga.us.prserv.net [165.87.234.18]
  4  nyor1br2-at-1-1-0-1.ny.us.prserv.net [165.87.230.38]
  5  nyor1sr3-ge-6-0-0-0.ny.us.prserv.net [165.87.28.183]
  6  jfk-brdr-01.inet.qwest.net [205.171.4.37]
  7  jfk-core-02.inet.qwest.net [205.171.30.17]
  8  jfk-core-01.inet.qwest.net [205.171.30.1]
  9  bos-core-02.inet.qwest.net [205.171.8.17]
 10  205.171.28.30
 11  63.145.0.14
 12  sloth.net.yale.edu [130.132.1.17]
 13  elsinore.cis.yale.edu [130.132.143.21]

Computing statistics for 325 seconds...
```

Exhibit 9. Pathping Initially Lists the Discovered Path to the Target and Then Displays the Duration of the Test Required to Gather and Compute Statistics

```
 10  205.171.28.30
 11  63.145.0.14
 12  sloth.net.yale.edu [130.132.1.17]
 13  elsinore.cis.yale.edu [130.132.143.21]

Computing statistics for 325 seconds...
             Source to Here   This Node/Link
Hop  RTT     Lost/Sent = Pct  Lost/Sent = Pct  Address
  0                                             passed [166.72.216.233]
                                0/ 100 =   0%   :
  1  352ms   0/ 100 =   0%     0/ 100 =   0%   204.146.251.94
                                0/ 100 =   0%   :
  2  389ms   0/ 100 =   0%     0/ 100 =   0%   atla1ar2-5-0-18.ga.us.prserv.net [165.87.132.178]
                                0/ 100 =   0%   :
  3  390ms   0/ 100 =   0%     0/ 100 =   0%   atla1br2-11-0-0.ga.us.prserv.net [165.87.234.18]
                                0/ 100 =   0%   :
  4  409ms   0/ 100 =   0%     0/ 100 =   0%   nyor1br2-at-1-1-0-1.ny.us.prserv.net [165.87.230.38]
                                0/ 100 =   0%   :
  5  441ms   0/ 100 =   0%     0/ 100 =   0%   nyor1sr3-ge-6-0-0-0.ny.us.prserv.net [165.87.28.183]
                                0/ 100 =   0%   :
  6  410ms   0/ 100 =   0%     0/ 100 =   0%   jfk-brdr-01.inet.qwest.net [205.171.4.37]
                                0/ 100 =   0%   :
  7  392ms   0/ 100 =   0%     0/ 100 =   0%   jfk-core-02.inet.qwest.net [205.171.30.17]
                                0/ 100 =   0%   :
  8  370ms   0/ 100 =   0%     0/ 100 =   0%   jfk-core-01.inet.qwest.net [205.171.30.1]
                                0/ 100 =   0%   :
  9  366ms   0/ 100 =   0%     0/ 100 =   0%   bos-core-02.inet.qwest.net [205.171.8.17]
                                0/ 100 =   0%   :
 10  370ms   0/ 100 =   0%     0/ 100 =   0%   205.171.28.30
                                0/ 100 =   0%   :
 11  366ms   0/ 100 =   0%     0/ 100 =   0%   63.145.0.14
                                0/ 100 =   0%   :
 12  389ms   0/ 100 =   0%     0/ 100 =   0%   sloth.net.yale.edu [130.132.1.17]
                                0/ 100 =   0%   :
 13  407ms   0/ 100 =   0%     0/ 100 =   0%   elsinore.cis.yale.edu [130.132.143.21]

Trace complete.

C:\WINDOWS>
C:\WINDOWS>
```

Exhibit 10.   A Summary of Round-Trip Delay and Lost Datagrams for Each Hop on the Route to the Target Address

Nslookup

Nslookup represents a utility tool that allows you to examine DNS server records as well as obtain information about an organization. Unfortunately, due to the proliferation of unscrupulous persons, many organizations disable the response to certain types of nslookup operations.

Overview

Exhibit 11 illustrates the use of nslookup with its Help command to display a list of program options. There are two methods by which you can use nslookup. You can enter the name of the program with a target address to obtain information about a domain, a method referred to as a command operation. As an alternative, you can enter the command nslookup by itself to enter the program's interactive mode of operation. Once in the interactive mode of operation, the prompt in the form of a greater than (>) sign is displayed, enabling you to enter commands. In Exhibit 11, we first entered the command nslookup by itself. This results in the display of the DNS server that will perform DNS operations you request and the IP address of the server. After this information was displayed, a greater than (>) sign was displayed to signify the program is in its interactive mode of operation. Entering *help* then resulted in the display of commands supported by nslookup.

```
MS-DOS Prompt - nslookup
C:\WINDOWS>nslookup
Default Server:  nscache.prserv.net
Address:  165.87.13.129

> help
Commands:   (identifiers are shown in uppercase, [] means optional)
NAME            - print info about the host/domain NAME using default server
NAME1 NAME2     - as above, but use NAME2 as server
help or ?       - print info on common commands
set OPTION      - set an option
    all             - print options, current server and host
    [no]debug       - print debugging information
    [no]d2          - print exhaustive debugging information
    [no]defname     - append domain name to each query
    [no]recurse     - ask for recursive answer to query
    [no]search      - use domain search list
    [no]vc          - always use a virtual circuit
    domain=NAME     - set default domain name to NAME
    srchlist=N1[/N2/.../N6] - set domain to N1 and search list to N1,N2, etc.
    root=NAME       - set root server to NAME
    retry=X         - set number of retries to X
    timeout=X       - set initial time-out interval to X seconds
    type=X          - set query type (ex. A,ANY,CNAME,MX,NS,PTR,SOA,SRV)
    querytype=X     - same as type
    class=X         - set query class (ex. IN (Internet), ANY)
    [no]msxfr       - use MS fast zone transfer
    ixfrver=X       - current version to use in IXFR transfer request
server NAME     - set default server to NAME, using current default server
lserver NAME    - set default server to NAME, using initial server
finger [USER]   - finger the optional NAME at the current default host
root            - set current default server to the root
ls [opt] DOMAIN [> FILE] - list addresses in DOMAIN (optional: output to FILE)
    -a          - list canonical names and aliases
    -d          - list all records
    -t TYPE     - list records of the given type (e.g. A,CNAME,MX,NS,PTR etc.)
view FILE       - sort an 'ls' output file and view it with pg
exit            - exit the program
>
```

Exhibit 11. The Commands Supported by the Nslookup Command

Operation

Exhibit 12 illustrates the use of nslookup in its command and interactive modes of operation. In the command mode, we used nslookup to access the DNS server at Yale University. In the interactive mode of operation after entering the domain yale.edu, we attempted to use the ls command with the -d option to list all records in Yale's DNS. From the last line in Exhibit 12, you can see that our query was refused. As previously mentioned, most organizations restrict *record surfing* as a security measure.

Now that we have examined nslookup, the operation and utilization of one additional utility program will be discussed. That program is ipconfig.

Ipconfig

To conclude this chapter, we will turn our attention to the use of the ipconfig utility program. This program represents a very helpful troubleshooting utility and it should be considered whenever you are experiencing a TCP/IP networking problem that appears to represent some type of configuration problem.

The ipconfig program is applicable to all versions of Windows other than Windows 95. If you are using Windows 95, the program is called winipcfg. In the remainder of this section, all examples of the operation and utilization of the program will use its more modern name.

```
MS-DOS Prompt - nslookup                                    _ □ ×

C:\WINDOWS>nslookup yale.edu
Server:   nscache.prserv.net
Address:  165.87.13.129

Non-authoritative answer:
Name:     yale.edu
Address:  128.36.236.12


C:\WINDOWS>nslookup
Default Server:  nscache.prserv.net
Address:  165.87.13.129

> yale.edu
Server:  nscache.prserv.net
Address:  165.87.13.129

Non-authoritative answer:
Name:     yale.edu
Address:  128.36.236.12

> ls -d yale.edu
[nscache.prserv.net]
*** Can't list domain yale.edu: Query refused
>
```

Exhibit 12.   Nslookup in Its Command and Interactive Modes of Operation

Overview

As its name implies, ipconfig provides configuration information. That configuration information can include data about the Windows IP configuration as
well as data concerning the LAN adapter used by the computer. Concerning
the latter, ipconfig is not restricted to providing information about the TCP/IP
protocol stack because it can even be used to display information about the
MAC address of your computer's LAN adapter.

Operation

The generation of a display screen indicating the options supported under
ipconfig is a bit different from the previously described Microsoft utility
programs. As noted in Exhibit 13 which illustrates the display of ipconfig
options, you need to enter the command followed by a forward slash (/) and
question mark (?) to generate a display of program options.

   In examining the entries in Exhibit 13, note that you can use the ipconfig
command not only to display information about your TCP/IP configuration,
but in addition to release the IP address for the specified adapter. The latter
provides a quick mechanism to remove an improper IP address setting. Also
note that the program supports a *renew* option that results in the renewal of
an IP address and a *flushdns* option, with the latter used to remove all DNS
resolved IP addresses and host names from cache memory. Due to the

```
MS-DOS Prompt                                                    _|□|×|
C:\WINDOWS>ipconfig /?                                              ▲

Windows 2000 IP Configuration


USAGE:
    ipconfig [/? | /all | /release [adapter] | /renew [adapter]
             | /flushdns | /registerdns
             | /showclassid adapter
             | /setclassid adapter [classidtoset] ]

    adapter    Full name or pattern with '*' and '?' to 'match',
               * matches any character, ? matches one character.
    Options
        /?              Display this help message.
        /all            Display full configuration information.
        /release        Release the IP address for the specified adapter.
        /renew          Renew the IP address for the specified adapter.
        /flushdns       Purges the DNS Resolver cache.
        /registerdns    Refreshes all DHCP leases and re-registers DNS names
        /displaydns     Display the contents of the DNS Resolver Cache.
        /showclassid    Displays all the dhcp class IDs allowed for adapter.
        /setclassid     Modifies the dhcp class id.

The default is to display only the IP address, subnet mask and
default gateway for each adapter bound to TCP/IP.

For Release and Renew, if no adapter name is specified, then the IP address
leases for all adapters bound to TCP/IP will be released or renewed.

For SetClassID, if no class id is specified, then the classid is removed.

Examples:
    > ipconfig                      ... Show information.
    > ipconfig /all                 ... Show detailed information
    > ipconfig /renew               ... renew all adapaters
    > ipconfig /renew EL*           ... renew adapters named EL....
    > ipconfig /release *ELINK?21*  ... release all matching adapters,
                                        eg. ELINK-21, myELELINKi21adapter.   ▼
```

Exhibit 13.   Information about the Use of the Ipconfig Command

complexity of the command options, the lower portion of the help screen includes five examples of its use.

The most popular use of the program is for the display of information, so we will turn our attention to this topic. In doing so, we will use the /all option which generates detailed information about the configuration of the computer we are using.

Utilization

In examining the use of ipconfig, I selected two operating scenarios. First, I used a computer connected via a LAN as a platform for the use of the ipconfig command. This was followed by the use of a computer with a modem connection to the Internet to obtain TCP/IP connectivity.

Exhibit 14 illustrates the use of ipconfig to display configuration information concerning a Windows 2000 platform configured to receive its IP address from a DHCP server. In examining Exhibit 14 you will notice that the display shows DHCP is enabled and the autoconfigured IP address assigned to the computer.

```
⌨ Command Prompt                                          ▓▓▓▓                    _|□|×|
                                                                                    ▲
E:\>ipconfig/all

Windows 2000 IP Configuration

        Host Name . . . . . . . . . . . . : gilbert-ne4a5au
        Primary DNS Suffix  . . . . . . . :
        Node Type . . . . . . . . . . . . : Broadcast
        IP Routing Enabled. . . . . . . . : No
        WINS Proxy Enabled. . . . . . . . : No

Ethernet adapter Local Area Connection:

        Connection-specific DNS Suffix  . :
        Description . . . . . . . . . . . : 3Com EtherLink XL 10/100 PCI NIC (3C
905-TX>
        Physical Address. . . . . . . . . : 00-60-08-33-23-29
        DHCP Enabled. . . . . . . . . . . : Yes
        Autoconfiguration Enabled . . . . : Yes
        Autoconfiguration IP Address. . . : 169.254.121.98
        Subnet Mask . . . . . . . . . . . : 255.255.0.0
        Default Gateway . . . . . . . . . :
        DNS Servers . . . . . . . . . . . : 205.131.174.1

E:\>_
                                                                                    ▼
```

Exhibit 14. Using Ipconfig with the /all Option

In addition to providing TCP/IP information, note that the use of ipconfig with the /all option also provides data about the LAN adapter. This information includes data about the type of LAN adapter and its MAC address. Thus, the use of ipconfig with the /all option can be used to obtain a considerable amount of configuration information.

In concluding this section, we will discuss obtaining information about a configuration obtained via a dial-up modem connection to the Internet. Exhibit 15 illustrates the use of ipconfig on my home computer after a dial-up Internet connection was established. Note that the term PPP represents an abbreviation for the Point to Point Protocol used to establish a serial TCP/IP dial-up connection.

In examining Exhibit 15, it should be mentioned that a dial-up connection to the Internet results in the temporary assignment of an IP address to a computer for the duration of the connection. It should also be noted that depending on the telephone number dialed, you may be routed to a different location within an ISP's network and receive an IP address from a different DHCP server. Similarly, your default gateway can be expected to change. In spite of the dynamics associated with dial-up modem access, you can still use ipconfig as a mechanism to note addresses assigned to your computer as a mechanism to report problems with information that may assist your ISP in resolving such problems. Similarly, if your organization operates dial-in servers to provide employees with a mechanism to access a corporate intranet, you can use ipconfig as a mechanism to obtain information about the connection as a tool for troubleshooting problems. Thus, ipconfig represents a utility program that covers a wide spectrum of possible uses.

```
MS-DOS Prompt                                              _ □ ×

C:\WINDOWS>ipconfig

Windows 2000 IP Configuration

PPP adapter Spinway.com_DO_NOT_EDIT:

        Connection-specific DNS Suffix  . :
        IP Address. . . . . . . . . . . : 12.64.42.149
        Subnet Mask . . . . . . . . . . : 255.255.255.255
        Default Gateway . . . . . . . . : 12.64.42.149

C:\WINDOWS>
```

Exhibit 15. Configuration Information Concerning a Dial-Up Modem Connection to the Internet

Index